FRAMING ROOFS

THE BEST OF
Fine Homebuilding

FRAMING ROOFS

THE BEST OF
Fine Homebuilding

The Taunton Press

Cover photo: Charles Miller

Back-cover photos: Charles Miller (top), James Docker (center), Robert Wedemeyer (bottom)

BOOKS & VIDEOS

for fellow enthusiasts

First printing: 1996
Second printing: 1998
Printed in the United States of America

A Fine Homebuilding Book

Fine Homebuilding® is a trademark of The Taunton Press, Inc., registered in the U.S. Patent and Trademark Office.

The Taunton Press, Inc.
63 South Main Street
P.O. Box 5506
email: tp@taunton.com

Distributed by Publisher Group West

Library of Congress Cataloging-in-Publication Data

Framing roofs : the best of Fine homebuilding.
    p. cm.
    Includes index.
    ISBN 1-56158-147-X
    1. Roofs — Design and construction.   2. Framing (Building).
  I. Taunton Press.   II. Fine homebuilding.
TH2393.F73   1996
694'.2 — dc20   96-7226
                                       CIP

CONTENTS

INTRODUCTION

I STOPPED off for a drink after work the other day and was chatting with Don, the bartender. He works as a roofer during the day so we often talk about construction. Don told me his father is a carpenter of the old school who can figure out anything with a framing square. The first time they framed a roof together, Don's dad asked him if he had a framing square, and Don replied, "Oh, you mean a corner checker?"

I suspect that on most job sites these days the once-proud framing square is relegated to the role of checking corners for square (when it's not being used to scrape ice from frosty lumber). Well, in this collection of articles from back issues of *Fine Homebuilding*, you'll learn what Don's dad already knows: how to calculate rafter lengths with a framing square; how to mark ridge cuts and bird's mouths; and how to find the common difference between jack rafters. You'll also learn a simple method for framing a valley, how to raise trusses safely, how to build dormers, and a whole lot more.

Written by builders from around the country, these articles keep alive the lore of the old-school carpenters, and some articles feature innovations, such as framing with a scientific calculator, that lay the groundwork for the new school.

—Kevin Ireton, editor

Roof Framing Simplified

This direct approach involves full-size layouts and stringing rafter lines

by Tom Law

There isn't a cut in roof framing that can't be calculated given a sharp pencil, a framing square and a head for math. But my 20 years in the trade have taught me that in some cases, the theoretical calculation of rafter angles and lengths is slower and leaves more room for error. While I think that it's important to understand the geometry of roof framing, the empirical method can save time and frustration, and contribute to your understanding of the process. I'm better at solving problems when I can grasp them—literally.

When I'm cutting a complicated roof and things get foggy, I use two techniques to help me produce rafters that fit the first time. I chalk lines on the plywood subfloor to represent a rafter pair in relation to its plates and ridge. This two-dimensional diagram is laid out full size. Pattern rafters can be tested right there on the job site. The other method I use is to deal directly with the components involved by getting up on the roof and measuring the relationships between the rafter to be cut and the existing plate and ridge with string and sliding bevel.

I used this method several years ago when I built a Y-shaped house. One wing was for the bedrooms and the other contained the kitchen, dining room and family room. The stem of the Y was the living room. The roof over the wings called for trusses, but the living-room rafters were exposed. The problem was in framing the intersection of the three roofs. These beams were big, long and expensive, and all the cuts would show. Had the house been a T shape, the valley rafters would have been a textbook case, and I could have found the information I needed in the rafter tables on my framing square. But since the intersection was 120° and not 90°, I had to find the angles by calculation or direct observation. I chose the latter.

Full-size layout—After setting the trusses, I chalked a full-size layout of the living-room common rafters and ridge on the subfloor below. The ridge beam was a 4x14, and the common rafters were 4x8s on 4-ft. centers. I decided to tackle the easiest steps first. This gave me time to think about the problem while reducing the parts in the puzzle.

The ridge beam went up first. I found its height by measuring on the full-size layout. The common rafters were then cut using patterns made from the layout. I nailed these in place

Tom Law is a builder in Davidsonville, Md.

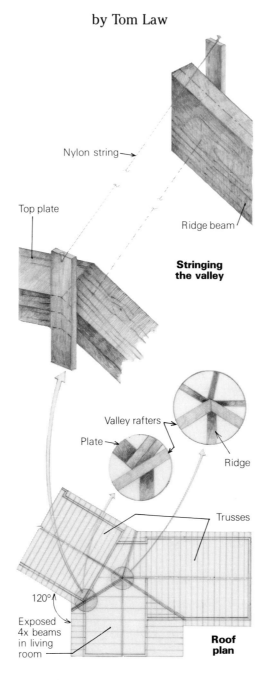

Nylon string

Top plate

Ridge beam

Stringing the valley

Valley rafters

Plate

Ridge

Trusses

120°

Exposed 4x beams in living room

Roof plan

starting at the outside wall, working toward the junction of the Y.

The valley rafter was next. Its location is shown above. Valley rafters are usually heftier than common rafters because they have a longer span, and have to carry the additional weight of the valley jacks. In this case, the valley rafter was a 4x10. Because it's deeper, to achieve the same height above the plate and ridge, its seat and ridge cuts also differed from those of the common rafters. I found it faster and easier to measure the actual distances than

to calculate imaginary ones. This way I could find the length of the rafter and the angles of the cuts without guesswork or error. This took me back up on the roof armed with a ball of nylon string, a sliding bevel and a level.

String lines—First, I tacked a scrap piece of wood vertically on the opposite side of the ridge beam from its intersection with the valley rafter. Then I tacked another block on the outside of the top plate where the valley rafter would sit. Between these sticks, I stretched nylon string at the height of the top of the rafter and along its imaginary center line. The string made it easy to visualize the actual rafter in place. For reassurance, I sighted across the rafter tops from the outside wall to check the alignment. I used the sliding bevel and level to find the angle of all the cuts, being careful not to distort the string.

Before I transferred the angles to the rafter stock, I made two templates (called layout tees) for marking out the ridge cut and rafter seats. I made one for the bottom and one for the top of the valley rafter. With some adjustments, they fit when the center line drawn on the pattern was in line with the string representing the center of the rafter. The tees also allow you to test-fit the bird's mouth to the plate before you carve up costly rafter stock. With these pattern pieces tacked in place, I measured the length of the valley directly. Then I transferred this length and the angles on the tees to the valley-rafter stock, and cut it to its finished dimensions. It fit perfectly the first time.

With the valley rafter in place, I turned to the valley jacks. First I made another pattern, this time of the ridge cut of the common rafter. I tacked it on the layout mark on the ridge and stretched the nylon line from it to the valley rafter, being careful to keep it exactly parallel to the common rafters. With the line simulating the top center line of the longest jack, I used the sliding bevel to find the angles of the plumb and side cuts. This time I transferred the angles directly onto the stock and cut it with a handsaw. Each shorter jack was worked in the same way, using the nylon line to find the location and length; the angles remained constant.

Laying out rafters with a framing square is something I do a lot. However, in situations that call for unusual intersections with compound angles, I spend my time dealing directly with the problem. This reduces confusion and allows me to concentrate on the work. □

Roof Framing Revisited

A graphic way to lay out rafters without using tables

by Scott McBride

When architect Stephen Tilly presented me with the drawings for a house he planned to build in Dobbs Ferry, N. Y., I flipped through the pages one at a time. I stopped when I got to the roof-framing plan. In addition to the roof's steep pitch (12-in-12) and substantial height (more than 40 ft. in some places), I would have to deal with a plan that included few corners of 90°, roof intersections of different pitches and a portion of a cone.

Irregular plan—When adjoining roof planes rise at the same pitch, their intersection (either a hip or valley) bisects the angle formed by their plates on the plan. For instance, a square outside corner on a plan calls for a hip whose run bisects the plates at 45°, as shown in the drawing below. This is why the skillsaw is set to a 45° bevel in most cases to cut hips. Similarly, square inside corners produce valleys lying at 45° to their adjoining ridges.

When plates join at an angle other than 90°, things get a bit more complicated. The Dobbs Ferry house contains a number of these irregular situations, including a large bay that angles off the master bedroom at 45°, and a wing that joins the main body of the house at 60°.

Theory—Conventional methods of roof framing involve the extensive use of rafter squares, but I decided to take a different approach. My study began by reading *Roof Framing* (Sterling Publishing Co., Two Park Ave., New York, N. Y. 10016) and *The Steel Square* (Drake Publishers, 381 Park. Ave. South, New York, N. Y. 10016), both by H. H. Siegele. These books are a hap-

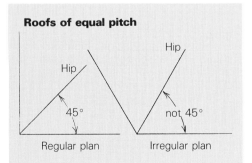

Roofs of equal pitch

Hip

Hip

45°

not 45°

Regular plan

Irregular plan

An irregular plan. With intersecting roof planes of the same pitch, the hip or valley bisects the angle formed by their plates in plan (drawing, above). On this house, right, the main wings join at 120°. As a result, the valley is an irregular one, since the angle at which it bisects the plates in plan is other than 45°.

Photos: Stephen Tilly

hazardly arranged series of magazine articles from the 1940s, and the illustrations are poorly reproduced in the current editions. Nevertheless, they contain the necessary applied geometry that is the essence of roof framing.

Understanding Siegele and the subject of irregular framing in general isn't easy. I spent hours rereading a page; in some cases, even a single cryptic paragraph. But when, after following his convoluted instructions, that weird cut I made on an 18-ft. double 2x10 valley rafter fit perfectly, it all seemed worthwhile.

Preparation—I began work on the house by making a ½-in. = 1-ft. scale model of the entire house frame (photo right). Lofted directly from

photographically enlarged working drawings, it allowed architect Tilly and me to analyze the downward transfer of the roof load, and to visualize the relationship of one rafter to another in three dimensions. Finally, the model proved helpful with the materials takeoff, since determining the rough length of hip, valley and jack rafters from a plan alone is difficult on a complex roof.

Once the last double top plate on the second story was nailed down, I began to deal with the roof in earnest. I drew two roof plans to use in the actual framing process. The first is a schematic (drawing, below) showing the position of plates, ridges, rafters, etc. This plan, like the model, gave me a general idea of how the roof

The structural model, above, shows just how complicated the roof is. The author had to deal with both an irregular plan and irregular pitches for some hip, valley and jack rafters, as well as a portion of a conical roof and dormers of several shapes. To keep on-site calculations to a minimum, he used a large developmental plan (drawing, below) that allowed angle takeoffs using a sliding T-bevel, and unit measurements using a pair of dividers.

Schematic roof plan
This plan was drawn to the same scale as the blueprints, and its purpose was to give the author an overall idea of how the entire roof went together—the exact number and location of all the plates, ridges, headers and rafters, both regular and irregular.

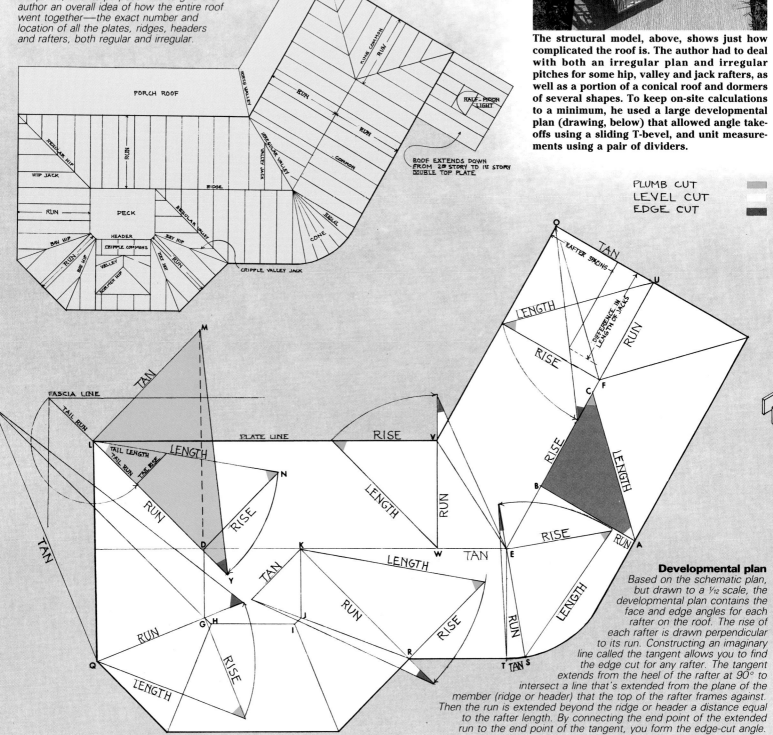

Developmental plan
Based on the schematic plan, but drawn to a 1/12 scale, the developmental plan contains the face and edge angles for each rafter on the roof. The rise of each rafter is drawn perpendicular to its run. Constructing an imaginary line called the tangent allows you to find the edge cut for any rafter. The tangent extends from the heel of the rafter at 90° to intersect a line that's extended from the plane of the member (ridge or header) that the top of the rafter frames against. Then the run is extended beyond the ridge or header a distance equal to the rafter length. By connecting the end point of the extended run to the end point of the tangent, you form the edge-cut angle.

would work, and I drew it to the same scale as the prints. On this plan, the common rafters are always oriented perpendicular to the plates, and the run, or horizontal distance traveled by the rafters, is the same on both sides of the roof.

I drew the second plan (drawing, bottom of facing page) at 1 in. = 1 ft.—four times as large as the first plan. This developmental plan contains the face and edge angles of each common, hip, valley and jack on the roof, making it possible for me to transfer them from the paper right onto the rafter stock with a sliding T-bevel. (For definitions of roof members and roofing cuts, see the drawing below.)

Also, the graphic constructions on this plan provide unit rafter lengths equal to 1/12 the extended rafter lengths since I used a 1-in. = 1-ft. scale. To get full-scale lengths, I adjust a large pair of dividers to the unit length on the plan and step off this distance twelve times along the measuring line on the rafter stock. This gives me the *extended length* (or unadjusted length) of each rafter. The extended length is the distance from the center of the ridge to the outside edge

of the top plate. To find the actual length of a common rafter, you must deduct one-half the thickness of the ridge and add the necessary amount to the tail for overhang at the eaves. I like this graphic system because it avoids math entirely—no rafter tables, no Pythagorean theorem. Also, the dividers are handier and more accurate for stepping off than the steel square.

Using the plan with commons—To draw the plan, I began with the development of a common rafter, which is seen in plan as line segment AB in the developmental drawing (facing page, bottom). The distance from A to B is the *run* of the rafter. It represents the distance across the attic floor from the outside edge of the plate to directly below the center of the ridge, and it forms the base of a right triangle. This triangle carries the geometric information you need to know about a rafter to cut it. The other leg of the triangle, line BC, is the *rise*. In actual three-dimensional space the rise goes from point B on the attic floor to point C, directly above B on top of the ridge. The hypotenuse

of this triangle, line AC, is the unadjusted length of the rafter. The angle opposite the run is the plumb-cut angle, and the angle opposite the rise is the seat-cut angle.

If you are thinking in three dimensions, you're probably confused. You can see that right triangle ABC is lying flat instead of standing in three dimensions as it would on a model. It is literally pushed over on one side, using AB as a kind of hinge. This is the key to understanding the way this developmental drawing expresses spatial relations using a flat piece of paper. The constructions for all of the rafters in the drawing employ the same basic transposition by bringing the rise and the rafter length down onto the paper, along with the run.

From paper to rafter stock—Cutting a common rafter from the developmental drawing requires taking the information graphically represented in the plan and applying it full scale to the rafter stock. For the common rafters represented by line AC, use a sliding T-bevel to transfer angle BCA onto the rafter stock. This will

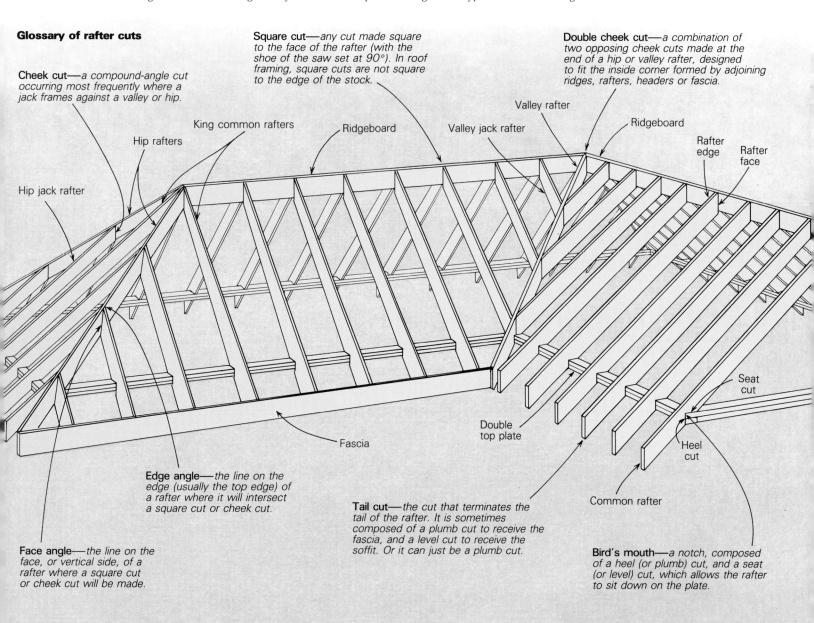

Glossary of rafter cuts

Cheek cut—a compound-angle cut occurring most frequently where a jack frames against a valley or hip.

Square cut—any cut made square to the face of the rafter (with the shoe of the saw set at 90°). In roof framing, square cuts are not square to the edge of the stock.

Double cheek cut—a combination of two opposing cheek cuts made at the end of a hip or valley rafter, designed to fit the inside corner formed by adjoining ridges, rafters, headers or fascia.

Hip jack rafter

Hip rafters

King common rafters

Ridgeboard

Valley jack rafter

Valley rafter

Ridgeboard

Rafter edge

Rafter face

Seat cut

Heel cut

Double top plate

Fascia

Common rafter

Edge angle—the line on the edge (usually the top edge) of a rafter where it will intersect a square cut or cheek cut.

Tail cut—the cut that terminates the tail of the rafter. It is sometimes composed of a plumb cut to receive the fascia, and a level cut to receive the soffit. Or it can just be a plumb cut.

Face angle—the line on the face, or vertical side, of a rafter where a square cut or cheek cut will be made.

Bird's mouth—a notch, composed of a heel (or plumb) cut, and a seat (or level) cut, which allows the rafter to sit down on the plate.

Drawing: Elizabeth Eaton

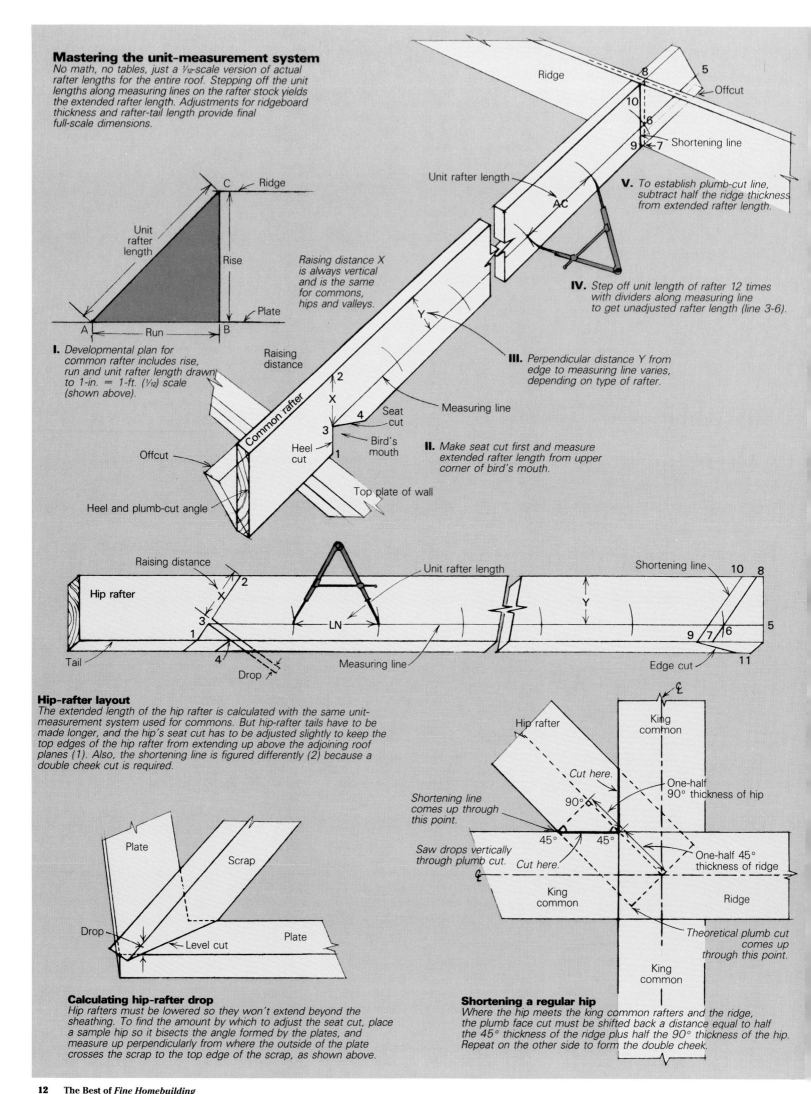

Mastering the unit-measurement system

No math, no tables, just a 1/12-scale version of actual rafter lengths for the entire roof. Stepping off the unit lengths along measuring lines on the rafter stock yields the extended rafter length. Adjustments for ridgeboard thickness and rafter-tail length provide final full-scale dimensions.

Ridge

8
5
Offcut
10
6
9 7 Shortening line

Unit rafter length

AC

V. *To establish plumb-cut line, subtract half the ridge thickness from extended rafter length.*

C — Ridge

Unit rafter length

Rise

Plate

A — Run — B

I. *Developmental plan for common rafter includes rise, run and unit rafter length drawn to 1-in. = 1-ft. (1/12) scale (shown above).*

Raising distance X is always vertical and is the same for commons, hips and valleys.

Raising distance

IV. *Step off unit length of rafter 12 times with dividers along measuring line to get unadjusted rafter length (line 3-6).*

III. *Perpendicular distance Y from edge to measuring line varies, depending on type of rafter.*

Common rafter

Y
2
X
3 1
4 Seat cut
Bird's mouth

Heel cut

Measuring line

II. *Make seat cut first and measure extended rafter length from upper corner of bird's mouth.*

Offcut

Heel and plumb-cut angle

Top plate of wall

Raising distance
Unit rafter length
Shortening line
10 8

Hip rafter
2
X
3
1
4
Drop
Tail

LN

Measuring line

Y

9 7 6 5
Edge cut 11

Hip-rafter layout

The extended length of the hip rafter is calculated with the same unit-measurement system used for commons. But hip-rafter tails have to be made longer, and the hip's seat cut has to be adjusted slightly to keep the top edges of the hip rafter from extending up above the adjoining roof planes (1). Also, the shortening line is figured differently (2) because a double cheek cut is required.

Plate
Scrap

Drop
Level cut
Plate

Calculating hip-rafter drop

Hip rafters must be lowered so they won't extend beyond the sheathing. To find the amount by which to adjust the seat cut, place a sample hip so it bisects the angle formed by the plates, and measure up perpendicularly from where the outside of the plate crosses the scrap to the top edge of the scrap, as shown above.

Hip rafter
King common

Cut here.
One-half 90° thickness of hip

Shortening line comes up through this point.
90°

45° 45°
One-half 45° thickness of ridge

Saw drops vertically through plumb cut.
Cut here.

King common
Ridge

Theoretical plumb cut comes up through this point.

King common

Shortening a regular hip

Where the hip meets the king common rafters and the ridge, the plumb face cut must be shifted back a distance equal to half the 45° thickness of the ridge plus half the 90° thickness of the hip. Repeat on the other side to form the double cheek.

give you a plumb line. Use it first to establish the heel cut 1-3 near the bottom of the rafter (drawing, facing page, top), making sure to leave enough stock below this line for the rafter tail.

Now, holding the blade of the T-bevel perpendicular to the heel cut, use angle BAC against the edge of the rafter to scribe the horizontal seat cut 3-4. I make the seat cut 3½ in. to equal the width of the rafter plate. Cutting away triangle 1-3-4 will establish the bird's mouth.

To establish a length on the rafter, you have to scribe a line to measure along. Starting at the inside corner of the bird's mouth (point 3), and proceeding parallel to the edge of the rafter, scribe the *measuring line* 3-5. Going back to the plan, set the dividers to the unit length of the rafter AC. Starting from point 3 again, step off this increment 12 times along the measuring line. This will bring you to point 6. Draw a plumb line 7-8 (it will be parallel to line 1-2) through this point. This represents the imaginary line down through the exact center of the ridge, and the distance from 3 to 6 is the unadjusted length of the rafter.

To make the actual face cut for the rafter where it bears against the ridge, draw line 9-10 (known as the *shortening line*) parallel to line 7-8. The shortening line is offset from the theoretical plumb cut by a horizontal distance equal to one-half the thickness of the ridge—in this case ¾ in.

Note that the top edge of the rafter is offset from the measuring line by the vertical distance 3-2 (equal to 6-8). This is what I call the raising distance (X). It represents the difference between the level at which the roof is calculated and laid out, and the tops of the actual rafters. Measuring lines on all rafters are offset from their respective edges by this same raising distance, as measured vertically along their respective plumb cuts. If all the planes of a roof are to meet smoothly, then this measurement must remain constant from rafter to rafter.

Raising the roof—After cutting all the common rafters, I cut temporary posts to support the ridge. I erected these at points D, E, and F (developmental drawing, p. 10) and toenailed ridge boards DE and EF in from above. Headers DG, GH, HI, IJ and JK were propped up at the same elevation as the ridge. There was a reason DG and KJ were not permitted to intersect directly with HI in a square corner: given the dimensions of the floor plan below, to have done so would have made the run of the canted bay roofs less than the run of the adjoining roofs. Since the rise of all roofs here is the same, and pitch is the function of run and rise, giving the deck a square corner would have given the bay roofs a slightly different pitch than their neighbors. This would have combined an irregular pitch condition with the existing irregular plan. Have mercy!

Regular hips—After spiking in all the commons, I proceeded to draw the regular hips (DL, OF, and PF) starting with DL on the developmental plan. I first drew the rise DN, perpendicular to the run (DL). Then I connected L to N to produce the unit length of the rafter as seen in profile. Note that the rafter length of the hip is

greater than that of the common, because its run is greater, even though the rise is the same.

In much the same way as I did with the common rafters, I used the T-bevel set to angle DNL from the plan to strike the heel cut on the rafter stock, again making sure to leave enough length on the tail for eventual trimming at the fascia (middle drawing, facing page). Hips and valleys require longer tails than commons and jacks, because although the rise of the hip tail is the same as the rise of the common tail, the run of the hip tail is greater. To calculate the length required for the hip tail, I drew in the fascia lines to scale on the plan parallel to the plate lines.

Extending the run of the hip until it intersects the fascia line produces the run of the tail. Set the dividers to this increment and swing them around 180° to a point along the hip run. Squaring up from this mark will establish the rise, and its intersection with the hip length establishes the end point for the tail length. Remember, though, this is a unit length, and will require being stepped off 12 times on the rafter.

The next task is to scribe the rafter stock with a measuring line from the inside corner of the seat cut to the top end of the rafter. It's parallel to the top edge of the rafter and offset from it by the same vertical raising distance established on the common rafters. To complete the bird's mouth on this hip, I drew a line through the intersection of the measuring line and the heel cut, using angle DLN on the T-bevel. This produced the seat cut. Well, almost.

Dropping the hip—A hip seat cut has to be *dropped*, which is the procedure of lowering the entire hip slightly in elevation by making a somewhat deeper seat cut. If the hip were not dropped, the corners on its top edge would protrude slightly beyond the adjoining roof planes, interfering with the installation of the sheathing.

To determine the amount of drop necessary, I took a piece of scrap the same thickness as the hip (drawing, facing page, bottom left) and made a level cut through it, using angle DLN. I placed this so that it bisected the 90° angle formed by the intersection of the adjoining plates. I then squared up from the level cut at the point where the outside of the plate crossed the face of the scrap. The distance from this point to the top edge of the scrap was the amount of drop. Accordingly, I moved up the theoretical seat cut a vertical distance equal to the drop, to arrive at the actual seat cut (middle drawing, facing page).

When stepping off a hip for length, you start from point 3 and repeat the unit hip-rafter length LN (obtained from the plan) 12 times. Using angle DNL, I struck a plumb line through the point 6. This is line 7-8, and represents the center of the ridge. Shifting this plumb line back horizontally a distance equal to half the 45° thickness of the ridge (1 1/16 in. for a 2x ridge) plus half the 90° thickness of the hip (¾ in. for a 2x hip) established the actual plumb face cut (shortening line), 9-10. The bottom right drawing on the facing page demonstrates why these reductions are necessary by showing hip plumb cuts.

To complete the hip rafter, the shortening line has to be laid out in the same relative position

on both sides of the stock, and cut with a skillsaw set at a 45° bevel from each side. This produces a *double cheek cut* (see the drawing on p. 11), which nuzzles into the corner formed by the adjoining faces of the king common rafters.

Irregular hips and valleys—After completing the regular hips, I was ready for a bigger challenge—the irregular parts of the plan. The easiest of the irregular rafters to cut (for reasons I will explain later) was the group of bay hip rafters typified by QG, so I started with those.

All cheek cuts, regular and irregular, fall into one of two categories: those that can be cut with a skillsaw (with a bevel of 45° or greater) and those that cannot (with a bevel less than 45°). This difference is determined by the angle formed between the run of the given rafter, and the ridge, header or other rafter against which it frames. For instance, the irregular hip rafter QG frames against header GH at a 67½° angle (112½° if measured from the opposite side of the rafter). As a result, it can be easily cut with a skillsaw, and in a sense is no more difficult to cut than a regular hip.

The first step is to develop the unit rise and the unit length of the rafter on the plan, using the same graphic procedure as for common rafters and the regular hips. Again, you will need to draw a measuring line parallel to the edge of the rafter stock, offsetting it by the same raising distance (X) used on all the other rafters. The raising distance is measured along the plumb cut, which is made using the angle shown in blue on the plan for QG. The seat cut for QG is made using the yellow-shaded angle from the plan. Remember that when using this graphic system, the plumb-cut angle is always opposite the run, and the seat-cut angle is always opposite the rise.

Dropping an irregular hip is done as on a regular hip, except that in this case, bisecting the irregular plate angle of this bay means orienting the scrap block at 67½° to the plates, not at 45°. The amount of drop is then measured up from the edge of the plate to the top edge of the scrap block as before.

Next, the unit rafter length from the drawing needs to be stepped off 12 times on the rafter stock to determine the unadjusted rafter length. However, the top plumb cut does not have to be shortened as the regular hip did because the face of header GH, against which QG frames, coincides with its theoretical layout line.

To make the plumb cut at the top end of the rafter, I set a 22½° bevel on my saw (the difference between 90° and 67½°), and cut away from the line. The resulting cheek fits against header GH, and a mirror image of this rafter fits header DG. Spiking these two together formed a V-notch at the top of the double rafter that straddled the 135° angle of the deck corner.

Adding the edge cut—The real challenge of an irregular plan comes when the bevel you are cutting is less than 45°. The bottom cut of valley KR is a good case in point. The bottom end of this rafter frames against the side of hip rafter JR, forming an angle in plan, KRJ, which is sharper than 45°. The solution is to lay out a

Facing page: The author began with the common rafters and the ridges and headers they frame against, and worked up to the more complex parts of the roof, like the dormer and bays (on the left in the top photo), the irregular valley that joins them to the common rafters (in the center of the photo). The bottom photo shows a convergence of two hipped bays, two dormers, and a doubled valley that connects this area to a run of commons.

face cut *and* an edge cut; and then to use either a handsaw or a chainsaw to make the cheek cut through both lines simultaneously (see the sidebar at right). It is in laying out the edge cut that things get a bit rough. I like to understand what I'm doing instead of just using a set of tables on a square, so I turned to Siegele again. His explanation relies on a line he calls the *tangent*, but its not the tangent you learned about in geometry class.

The tangent—This line is used in graphic constructions to find the edge bevel of roof members, roof sheathing, hopper boards, and other compound-angle cuts. Unfortunately, at no point in his books does Siegele give a concise definition of tangent. It's best understood in context, but I've assigned my own definition to the word just to introduce it. Basically, the tangent of a given rafter is an imaginary horizontal line, perpendicular in plan to the run of the rafter, extending from the heel of the rafter to the plane containing the vertical face of that member (ridge, header, or other rafter) against which the cheek cut of the rafter frames. Whew.

As an example, refer to the regular hip rafter LN, seen from above as LD (drawing, below). To

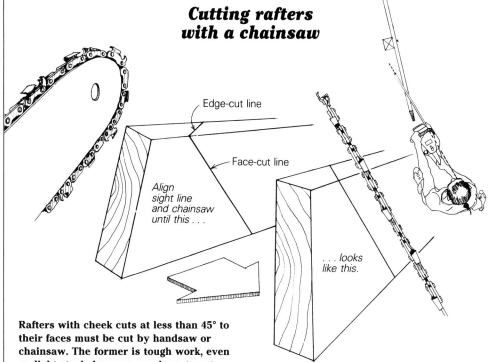

Cutting rafters with a chainsaw

Edge-cut line

Face-cut line

Align sight line and chainsaw until this . . .

. . . looks like this.

Rafters with cheek cuts at less than 45° to their faces must be cut by handsaw or chainsaw. The former is tough work, even on light stock, because you have to cut through the wood at a miter *and* a bevel. This means cutting a cross section greater than the area of the same stock cut square.

To cut rafters with a chainsaw, you must line up the face cut and the edge cut with your eye so they appear as a single line, and then introduce the bar of the chainsaw so that it also lines up along this plane of vision. When everything coincides, you can make the cut.

This all sounds simple enough until you consider that holding the saw as I just described violates the first rule of chainsaw safety: Keep your face and shoulder out of the plane of cut, because that is where the saw will go if it kicks back. But lining up like this is the only way

I know to produce a cheek cut that fits precisely. As a result, I use extra caution when I'm cutting this way. I spike the rafter stock to a post or some other solid object, leaving the end to be cut sticking well out in mid-air. Then I check the stock carefully for nails. I give myself plenty of working room, especially behind me. I keep my arms firmly locked and my legs spread, and drop the saw smoothly through the cut, without twisting. The chain must be kept sharp, well lubricated, and fairly tight on the bar; ripping chain works best.

After sawing, you'll probably want to smooth out the cut you just made with a plane. I use the 6-in. wide Makita 1805B and get excellent results. —S. M.

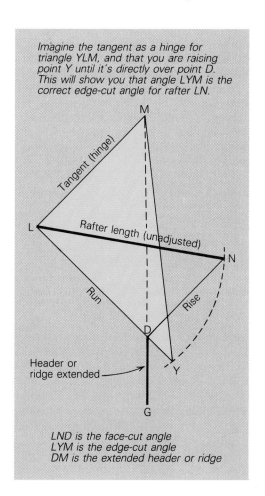

Imagine the tangent as a hinge for triangle YLM, and that you are raising point Y until it's directly over point D. This will show you that angle LYM is the correct edge-cut angle for rafter LN.

M

Tangent (hinge)

L

Rafter length (unadjusted)

N

Run

Rise

D

Y

Header or ridge extended

G

LND is the face-cut angle
LYM is the edge-cut angle
DM is the extended header or ridge

construct the tangent for this hip (LM), I have to draw a line perpendicular to the run of the hip LD that emanates from the heel of rafter LN. The tangent ends at point M, where it intersects the face of extended header DG. This imaginary plane is seen from above as a dotted line—DM—which extends from line DG.

To find the angle for the edge cut, I set the steel point of my compass at L and the pencil point at N and swing an arc down to intersect the extended line LD at Y. Now the extended run LY equals the rafter length. By connecting M to Y, I make angle MYL, which is the edge cut for the hip where it frames against header DG.

Why it works—I'll attempt to explain why this geometric construction works. Imagine that tangent LM, magnified now to full scale, is a sort of giant hinge. Slowly, triangle LYM (colored green in the drawing at left) starts to rise upward, led by point Y, while line LM remains anchored where it is. Your perspective on this is looking straight down.

When line MY coincides in your vision with the plane represented by line DM, and point Y coincides exactly with point D, the plane LYM is

inclined at the right pitch for the hip LN. Only at this pitch is the rafter length LY seen as equal in length to the run LD.

With plane LYM in this angled position, imagine a T-bevel applied with its stock along LY and its blade along YM. Now imagine a hip rafter, with its seat cut already made, raised up into position so that its top edge runs right along the raised-up line LY. If you survived these visualizing gymnastics, you'll begin to see why the angle LYM on the T-bevel gives the correct edge cut on the hip.

Here is another way of describing this concept: Construct a right triangle with legs equal to the tangent and rafter length—the angle opposite the tangent gives the edge cut.

For regular hips and valleys, the tangent is equal to the run. So by substituting the word run for the word tangent, we get a useful corollary: In all regular roof framing, use the unit run of the rafter on one arm of the square, and the unit length of the rafter on the other—the latter gives the edge cut. □

Scott McBride is a carpenter and contractor in Irvington-on-Hudson, N. Y.

Framing Gable Ends

You can use a story pole and a level to lay out accurate cuts without a tape measure

by Jim Thompson

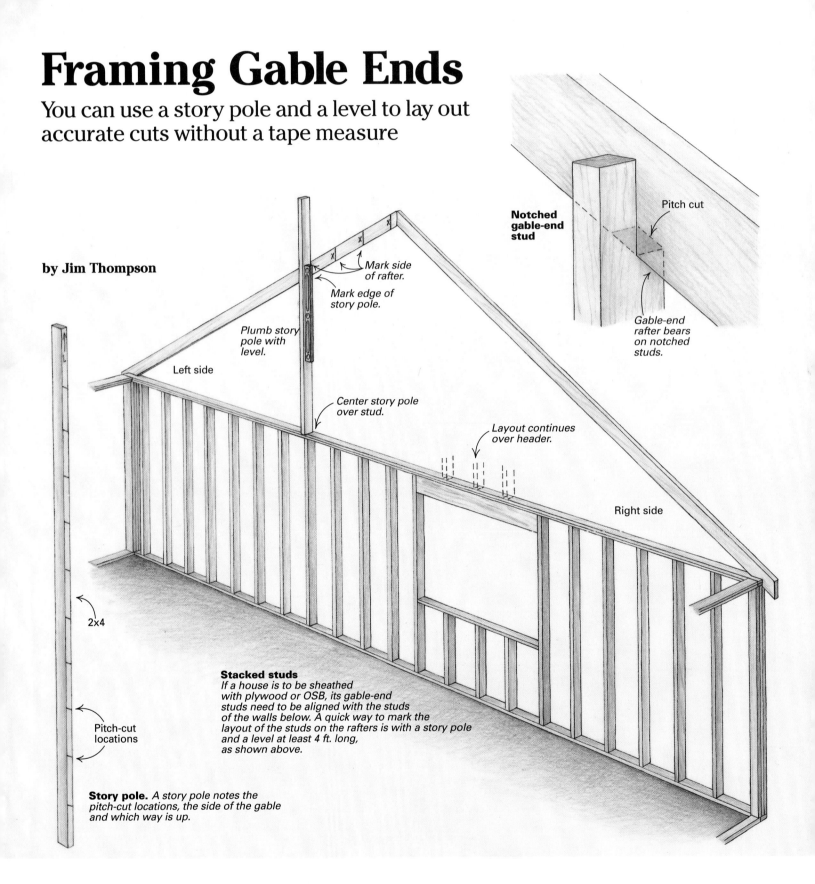

Notched gable-end stud

Pitch cut

Gable-end rafter bears on notched studs.

Mark side of rafter.

Mark edge of story pole.

Plumb story pole with level.

Left side

Center story pole over stud.

Layout continues over header.

Right side

2x4

Pitch-cut locations

Stacked studs
If a house is to be sheathed with plywood or OSB, its gable-end studs need to be aligned with the studs of the walls below. A quick way to mark the layout of the studs on the rafters is with a story pole and a level at least 4 ft. long, as shown above.

Story pole. *A story pole notes the pitch-cut locations, the side of the gable and which way is up.*

The first time I had to frame and sheathe the gable end of a house, I did a miserable job. The carpenter I was working for had no system for laying out the studs, which are all of different lengths and have angled cuts on their tops to match the roof pitch. He just said fill in the end between the last rafter and the plate below so that we could put sheathing on it. Then he took off. I did a lot of cutting and fitting and recutting and measuring. The job seemed to take forever.

A few years later another carpenter taught me the method I'm going to explain here. It may look confusing at first, but once you get it, you'll have a gable end framed in less time than it takes to read this article.

The studs are lined up over the wall studs below (called stacking) and notched around the gable-end rafter (drawings above). The important part of the notch is the pitch cut, which is on the same angle as the rafter (right drawings,

facing page). When the studs are cut and installed this way, they flush up to both the interior and exterior of the framing. I like the solid feel of gable walls that are framed with this method.

Before installing the gable-end studs, a little preparation is required. You should place a couple of sheets of plywood on the ceiling joists for a work area. On this work area you need a circular saw, a level that's at least 4 ft. long, a framing square and enough studs and sheathing to cover

Drawings: Christopher Clapp

the gable. It's also helpful to leave out the next-to-last rafters to give yourself adequate room to work. I only forgot that trick once.

Story pole—Begin the job by selecting a straight stud that is long enough to extend above the top of the rafter at your highest stud. This is your story pole, and it eventually will have the height of the notches for all the gable studs marked on it. Draw a big R on one edge to indicate the right side of the gable, and a big L on the opposite edge to record the left side. Then put an arrow on one edge to show which way is up.

Begin marking at the high end of the gable by putting the bottom of the story pole directly over the stud nearest the peak. As shown in the drawing on the facing page, the story pole is held against the rafter at the top and plumbed with the level. Next make a mark on the rafter to show the uphill side of the story pole and another mark on the edge of the story pole that records the bottom of the rafter. With the marks made, pull the story pole away from the rafter and draw an X on the downhill side of the line on the rafter to show where the stud will go.

Repeat this procedure for all of the studs on one side of the gable, then turn the story pole around and do the same for the studs along the other side. When you're finished, your rafters should be marked for stud layout, and your story pole should have similar markings on both sides that show the positions of the pitch cuts (left drawing, facing page).

Making the cuts—Now spread out enough studs to do the right side of the gable. Lay the studs on edge, snug them up to the story pole and use a framing square and a sharp pencil to transfer the story-pole marks to the studs (left drawing, this page).

The marks on the edges of the studs are registration points for your first cut. Arrange the studs so that the marks line up (middle drawing, this page) and run the circular saw down the line with the depth set equal to the thickness of the rafter (typically 1½ in.).

The next step is to cut off the tops of the studs so that they don't extend above the rafters. I leave 3½ in. above the pitch cut. That's enough material to anchor a nail—any more wastes wood. The wide side of my saw's base (Skil 77) is 3½ in. from the edge to the blade. So I eliminate measuring and use the saw's base to tell me how far from the pitch cut to make my crosscut. As I make my crosscuts, I set aside the offcuts. They will be the studs for the left side of the gable.

Now I can cut out the waste to make the notches. Because the narrow side of my saw's base is 1½ in. wide, I use the edge of the base to line up my cut (bottom right drawing, this page). Resist the temptation to hold the stud in one hand while operating the saw with the other hand. Instead, tack or clamp the stud to a work surface to hold the stud steady during the cut.

When you flip over the story pole and line up the offcuts for marking the left side of the gable, you'll notice that most of the offcuts are usable. In this example, the job requires only two additional studs, and the story pole can be used for

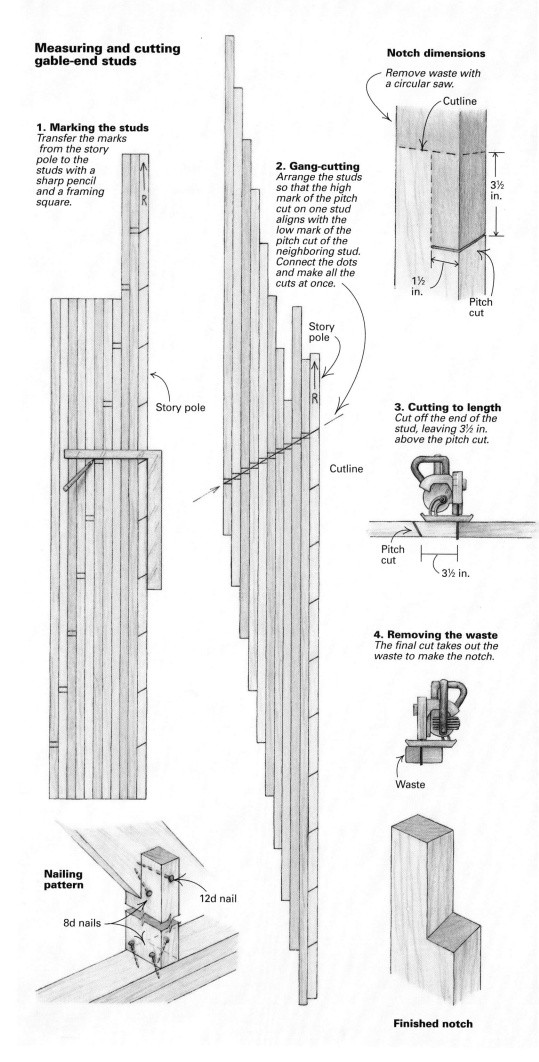

Measuring and cutting gable-end studs

1. Marking the studs
Transfer the marks from the story pole to the studs with a sharp pencil and a framing square.

Story pole

2. Gang-cutting
Arrange the studs so that the high mark of the pitch cut on one stud aligns with the low mark of the pitch cut of the neighboring stud. Connect the dots and make all the cuts at once.

Story pole

Cutline

Nailing pattern

12d nail

8d nails

Notch dimensions

Remove waste with a circular saw.

Cutline

3½ in.

1½ in.

Pitch cut

3. Cutting to length
Cut off the end of the stud, leaving 3½ in. above the pitch cut.

Pitch cut

3½ in.

4. Removing the waste
The final cut takes out the waste to make the notch.

Waste

Finished notch

one of them. This economy of material will vary with the pitch of the roof, the layout of the wall below and the length of the wall, but I've found the method to generate very little waste.

Stud and sheathing installation—To install a stud, set its bottom on the plate and align it with the stud below. (I prefer to start with the tallest stud and work toward the eaves, but it doesn't really make any difference.) Now slide the notch under the rafter until it is just snug. Be careful not to push too hard because you can easily put a hump in the rafter with the first stud. Then none of the studs will fit correctly.

Toenail the bottom of the stud to the top plate with two 8d nails on one side and a single 8d nail on the other side. Nail the top with a 12d through the top of the stud into the rafter. Then add an 8d toenail into the bottom of the rafter from the up-hill side of the stud.

I prefer to sheathe a gable end from inside the building, so I rip my sheathing into 2 ft. by 8 ft. pieces. That's because it's easy to lean out and nail off the bottom of a 2-ft. wide piece of sheathing, but it's tough to reach the bottom of a 4-ft. wide piece. (In areas subject to earthquakes or high winds, this method may not fulfill shear-wall requirements.)

I start the first piece of sheathing so that its edge lands on the center of a stud that's on my 16-in. o. c. layout. Not all studs are placed at layout; for example, a gable-end vent may require studs to fall outside the layout.

Save the cutoffs from the pieces of sheathing that project above the rafters. You can use them for starting or finishing a course because the cutoffs already include the pitch cut. When your sheathing gets too high to reach over, nail a 2x4 across the gable-end studs to make a place to stand. Use two 16d nails in each stud for this piece. When you're finished, leave the 2x in place because it helps stiffen the gable end.

But what about...—Here are answers to the typical questions I get about this method. First, if your centers are regular, and the slope of the roof is constant, why not just mark the first stud, get the common difference and mark the studs that way? Some folks prefer that technique (see sidebar, right), but I have found that it really takes no time at all to mark the story pole, which allows me to lay out the stud position on the rafter at the same time. Having the rafter laid out helps avoid bowing it upward if the cut is a little off, and the stud is too tall or too short.

Second, why not lay out the right and left halves from the same marks? Unless your layout below is exactly centered on the peak, which I've never found to be the case, it won't work.

Probably the most important things to remember while using this method are the orientation of the story pole while transferring the marks and which direction to slide the studs when you're preparing to make the pitch cuts. You simply have to remember which end is up. □

Former framing carpenter Jim Thompson is now a structural engineer with McCormac Engineering Associates in Ellicott City, Md.

Cutting gable-end studs using the common difference

by Elmer Griggs

Most roofs I have worked on over my career as a rafter cutter have been gable roofs. And most of the houses I've worked on here in southern California have been finished with stucco—not plywood covered with siding as in other parts of the country. Instead of relying on plywood or OSB for diagonal braces, builders here use 1x6s that are let into the studs to brace the walls. As a consequence, the studs on a gable end don't have to line up with the studs in the wall below to continue the layout for sheathing. The gable-end studs simply have to end up on 16-in. centers to provide ample backing for the stucco.

The method I'll describe here is used by a production framer to determine the lengths and the pitch cuts of the gable-end studs by means of the common difference, which is the consistent incremental change in length of equally spaced studs under a rafter. With this method, I make a minimum number of cuts, and I can work on the ground next to the lumber pile.

Roof pitch—Designers typically show the pitch of a roof by drawing a right triangle on the elevation or the roof plan (top drawing, facing page). The base of the triangle represents 12 horizontal inches (the run), and the vertical leg of the triangle shows the amount the roof rises in that horizontal foot. The hypotenuse of the triangle therefore represents the pitch of the roof. This example has a 6-in-12 pitch, and the run of the roof is 15 ft.

To calculate the lengths of the gable studs, first find the length of the longest stud. Here's the equation. Multiply the pitch (6 in.) times the run (15). The answer is 90 in. (top drawing, facing page). A 90-in. stud would be necessary to support the rafter at its highest point. But there's a curveball. A gable end typically has a vent in it, so the studs on both sides of it must be moved over a bit so that they don't interfere with the vent. To allow for the vent, I deduct half the pitch from the longest stud. That reduces the stud by 3 in. to 87 in. long, and moves the stud over 6 in.

What's the difference?—The centerlines of gable studs are normally spaced 16 in., which is equal to 1⅓ ft. To find the difference in length between adjacent studs, multiply 1⅓ by the pitch. In my example the equation is 1⅓ x 6 = 8-in. difference between studs.

This gable end needs 11, 8-ft. 2x4s. Two studs are going to come out of each one of these 2x4s—one for the right side of the gable and one for the left side. The studs will be mirror images of each other.

On a flat work surface, place the 2x4s on edge with their ends square to one another

(middle drawing, facing page). Measure 87 in., and mark the 2x4 closest to you at that point with a pencil. Then mark each successive 2x4 8 in. shorter.

Cutting the rafters—A gable-end stud tucks under the rafter, so its end has to be cut at the same pitch as the rafter. There are two quick ways to find the correct angle. With a framing square, align the edge of a 2x4 with the 6-in. mark on the tongue of the square and the 12-in. mark on the blade (bottom right drawing, facing page). Mark the angle with a pencil. Set the table of your circular saw on the face of the 2x4 and adjust the angle of the table so that the blade matches the pencil line. Or you can look in a book on rafter cutting to see what angle a 6-in-12 pitch equals. The answer is 26½°. The book I use is *Full Length Roof Framer* (A. F. J. Reichers, P. O. Box 405, Palo Alto, Calif. 94302).

When you cut a stud with the saw set at an angle, the resulting mitered end gives the stud a long point and a short point (bottom left drawing, facing page). I make my measurements to the long points. And when I cut any material on a miter, I cut from the side that will allow my mark to be the long point of the cut.

To finish marking the gable studs, measure the length of the shortest stud. In my example, it's 7 in. long. As shown in the drawing, add that length to the 2x4 that has the longest stud laid out on it. The long point of the shortest stud will correspond to the short point of the longest stud.

With a framing square aligned on the mark that defines the bottom of the shortest stud, mark the waste cut across the rack of 2x4s with a pencil (middle drawing, facing page). Make this cut with the saw set for a square cut. Now set the saw table at 26½° and cut each 2x4 at the cutline marks. When you're finished, hold up the two shortest studs. They should be the same length.

Nailing—I start with the longest studs, and I use a level to make sure they're plumb. The studs should just touch the bottom of the rafter—don't wedge them in place, or you'll put a hump in the rafter.

After putting in the longest stud on one side of a gable, take one that will fit about in the middle of the rafter. Look down the rafter to see if it has a hump or a sag in it. Move the stud up or down to take out any hump or sag. Don't drive the nails through the stud into the rafter. Instead, toenail a couple of 16d nails through the outside face of the rafter into the tops of the studs (photo facing page).

Once you've got the middle stud in, fill in the remaining ones. Use the level to keep the long studs plumb. It's pretty easy to plumb the short ones by eye.

—Elmer Griggs is a former carpentry instructor for Ventura County and Los Angeles County apprenticeship committees. He lives in Reseda, Calif.

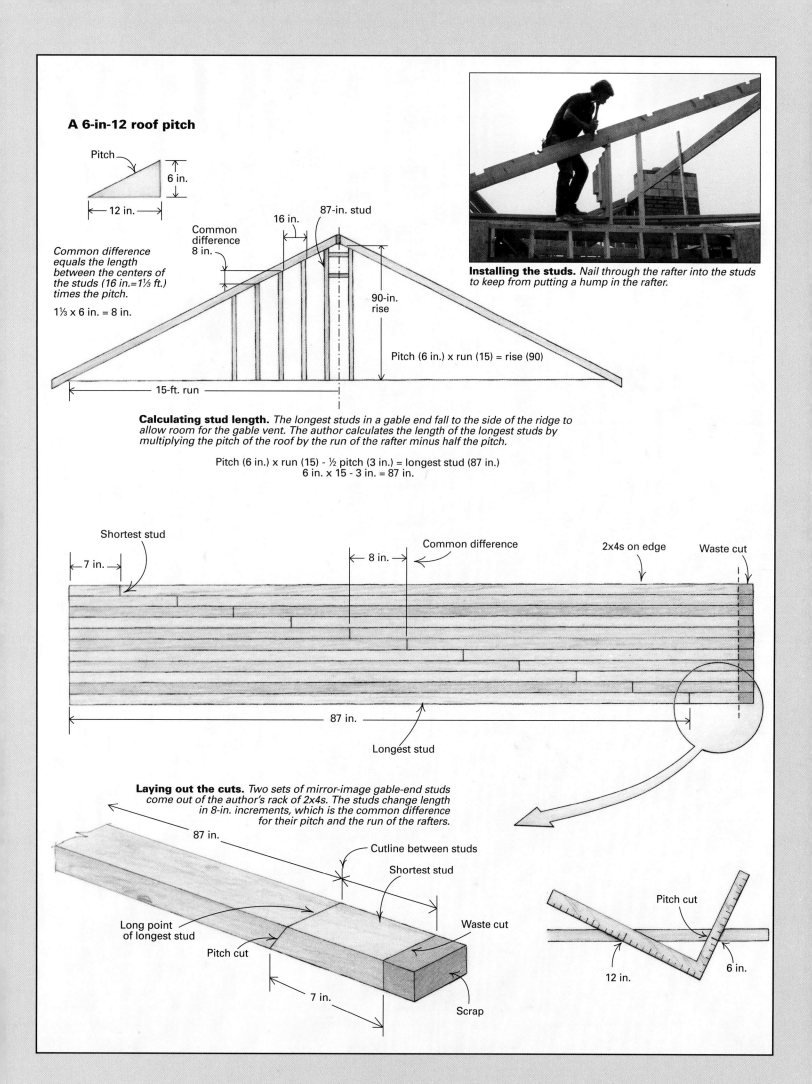

A 6-in-12 roof pitch

Pitch

6 in.

12 in.

Common difference equals the length between the centers of the studs (16 in.=1⅓ ft.) times the pitch.

1⅓ x 6 in. = 8 in.

Common difference 8 in.

16 in.

87-in. stud

90-in. rise

Pitch (6 in.) x run (15) = rise (90)

15-ft. run

Installing the studs. *Nail through the rafter into the studs to keep from putting a hump in the rafter.*

Calculating stud length. *The longest studs in a gable end fall to the side of the ridge to allow room for the gable vent. The author calculates the length of the longest studs by multiplying the pitch of the roof by the run of the rafter minus half the pitch.*

Pitch (6 in.) x run (15) - ½ pitch (3 in.) = longest stud (87 in.)
6 in. x 15 - 3 in. = 87 in.

Shortest stud

7 in.

Common difference

8 in.

2x4s on edge

Waste cut

87 in.

Longest stud

Laying out the cuts. *Two sets of mirror-image gable-end studs come out of the author's rack of 2x4s. The studs change length in 8-in. increments, which is the common difference for their pitch and the run of the rafters.*

87 in.

Cutline between studs

Shortest stud

Long point of longest stud

Pitch cut

Waste cut

7 in.

Scrap

Pitch cut

12 in.

6 in.

With its cross-gabled roof and decorative vergeboards, this cottage complements the Gothic home to which it's appended. Photo by Kevin Ireton.

Framing a Cross-Gable Roof

One good valley rafter supports another

by Scott McBride

In 1851 a German immigrant named Henry Kattenhorn owned a thriving sugar refinery in the riverfront village of Hastings-on-Hudson, New York. Deciding that his four superintendents and their families should share in his prosperity, Kattenhorn built cottages for them on a bluff overlooking the river. Bedecked with finials, decorative chimneys and gaily sawn vergeboards, these small, cozy houses were prime examples of Gothic Revival architecture.

Just a year before the Kattenhorn cottages were built, Andrew Jackson Downing, the leading exponent of the Gothic Revival style, had published *The Architecture of Country Houses*. In his book, Downing had inveighed "an excess of fanciful and flowing ornaments of a cardboard character," but the country carpenters who adapted the style from readily available pattern books were hard to restrain—lumber was cheap, the steam-driven jigsaw had been invented, and the sky was the limit.

Besides gingerbread, another hallmark of the Gothic Revival style was the cross-gable roof. Downing also tried to temper the proliferation of gables, lamenting that "some uneducated builders...have so overdone the matter, that, turn to which side of their houses we will, nothing but gables salutes our eyes." But the "cocked-hat cottage," as Downing called small dwellings with multiple gables, was precisely the form

chosen for a recent addition to one of the Kattenhorn cottages (photo above).

When Judy Seixas approached architect Stephen Tilly about adding a semi-detached bedroom suite to the back of her house, she was adamant that the design be strictly in keeping with the Gothic Revival style. Tilly and chief designer Laurel Rech came up with a simple cross-gable rectangle for the addition. An existing flat-roofed screen porch would be enclosed to house a bathroom, the utility room and an entrance foyer. The converted porch would also link the bedroom suite to the existing house. I was hired to build the addition, the trickiest part of which turned out to be framing the cross-gable roof.

Blind valleys—My crew framed partitions in the former screen porch while the foundation for the new addition was being built. As the blockwork was finished and floor framing began, I retired to a shady spot on the driveway to lay out and cut the principal roof members.

In a pure cross-gable roof, two ridges—both at the same elevation—intersect at 90°. All four valleys formed by the intersection converge at a central peak. Our addition would be a modified version, insofar as there would be a higher continuous ridge and a slightly lower ridge broken by the intervening higher gable. It could be called a gable with two dormers, except that I

think of dormers as being subordinate in size to a main roof. The similar size of all four gables on this roof makes them more or less equal partners in the deal.

I have seen cross-gable roofs in Victorian houses where a lower ridge flies right through the attic space under a higher ridge. But because our addition was to have a finished cathedral ceiling, I broke the lower ridge into two discontinuous sections. I considered supporting these lower ridges by hanging their inboard ends on headers framed between the common rafters of the higher gable, which is how I frame gable dormers. But the lower gables in this case were so broad that we would have needed a 13-ft. header to span the distance, which would have been an impractical arrangement.

We resorted to a supporting valley, or a blind valley. For each of the lower gable roofs, one valley rafter would run from the wall plate to the main ridge (photo facing page); this is the blind valley. The other valley would be shorter and would intersect the blind valley. This intersection marks the terminus of the lower ridge.

Because the addition's plan was symmetrical, it didn't matter which valley of a pair would run through to the ridge. But I did decide to make the blind valleys from opposing sides of the roof come together at the same point on the ridge. This way, any force exerted on the ridge by one

Blind valley. *Two factors complicated the framing of this roof: One pair of gables is lower than the other; and the room below gets a cathedral ceiling. The solution was to run one valley rafter on each side of the roof through to the main ridge. This is called a blind valley, and it carries the shorter valley and the lower ridge. The framing plan below shows a bird's-eye view of these parts.*

Framing plan

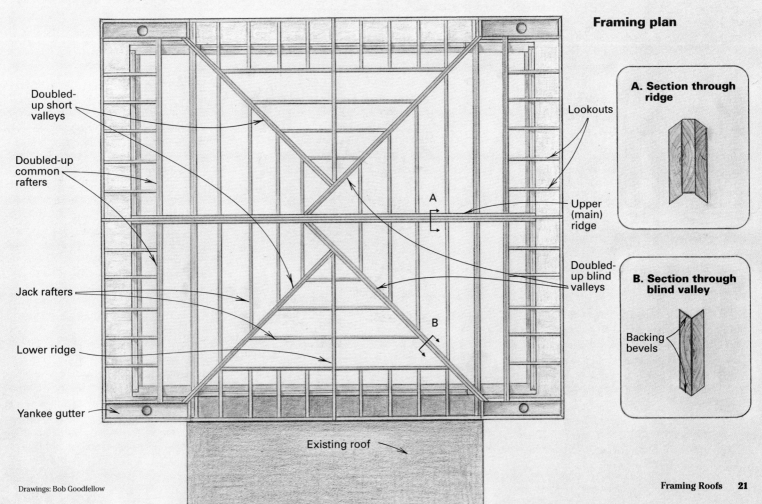

Doubled-up short valleys

Doubled-up common rafters

Jack rafters

Lower ridge

Yankee gutter

Lookouts

Upper (main) ridge

Doubled-up blind valleys

A. Section through ridge

B. Section through blind valley

Backing bevels

Existing roof

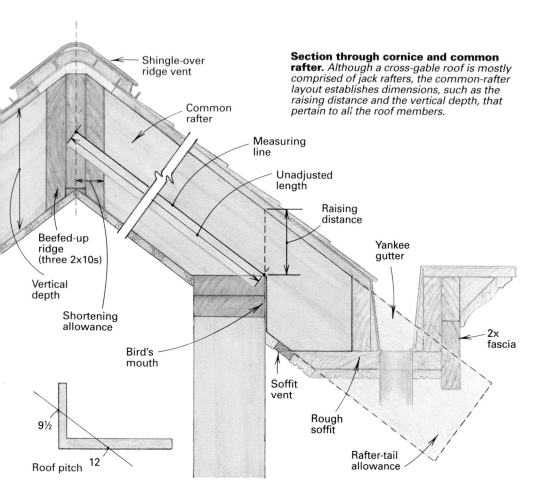

Section through cornice and common rafter. *Although a cross-gable roof is mostly comprised of jack rafters, the common-rafter layout establishes dimensions, such as the raising distance and the vertical depth, that pertain to all the roof members.*

Labels in figure:
- Shingle-over ridge vent
- Common rafter
- Measuring line
- Unadjusted length
- Raising distance
- Yankee gutter
- Beefed-up ridge (three 2x10s)
- Vertical depth
- Shortening allowance
- Bird's mouth
- 2x fascia
- Soffit vent
- Rough soffit
- Rafter-tail allowance
- 9½
- 12
- Roof pitch

blind valley would be canceled out by the force of the opposing blind valley.

Initially, we didn't want collar ties piercing the cathedral ceiling (although we added a pair, which I'll tell you about later). To compensate for the structural loss of the collar ties and to support the weight of the valleys, we beefed up the main ridge. Three 2x10s spiked together became a structural ridge beam. Because of the girth of the ridge, I beveled the top and bottom edges of the two outside 2x10s so that they wouldn't interfere with roof and ceiling planes.

The ridge beam was the first roof member to go up. We supported it on the end walls of the higher gable and put a temporary post under the spot where the blind valleys would meet.

Common-rafter layout—Because the ridges were at right angles to each other, and the roofs were the same pitch (9½-in-12), I was dealing with regular roof framing, meaning that the compound edge bevels on all my valley and jack rafters would be cut with my circular saw set at 45°. Knowing this, I decided to forego the graphic-development method I use to lay out complex, irregular roofs and resorted to more direct numerical methods.

On a clean sheet of plywood, I laid out a full-scale section of the cornice (drawing above). Next I drew in the top edge of the 2x8 common rafter and the measuring line, which is parallel to both edges and originates at the *outside* corner of the plate. The distance along an imaginary plumb line reaching from the measuring line to the top edge of the rafter is the *raising distance*—a key measurement that would remain constant for all the commons, the valleys and the ridges in the roof frame.

To figure the rafter-tail length, I referred to the blueprints. The architect had furnished me with a wall section showing a copper-lined Yankee gutter, which was to be recessed into the roof at the four short sections of eaves located at each corner of the addition. I couldn't envision how this cornice would return into the vergeboard of the higher gable at one end or how the valley flowing into the gutter would be resolved at the other end. I decided to play it safe by letting the valleys and the commons run long by a generous amount, figuring I'd trim them when I could see things in three dimensions.

After drawing the cornice section, I had the information I needed and could then transfer that information to the rafter stock. Laying a piece of rafter stock in front of me, I scribed my measuring line down its length, offset from the rafter's top edge by the raising distance noted earlier. From the end of the 2x8, I measured up 2 ft. for the rafter-tail allowance and drew a plumb line. Then I drew a level seat cut through the intersection of the plumb cut and the measuring line. I now had my bird's mouth. From the corner of the bird's mouth, I measured the unadjusted length of the rafter along the measuring line.

If the rise of the roof had been a whole number, such as 5-in-12, I could have found the rafter's length in a rafter table such as the one found stamped on the blade of my framing square. But because the pitch was 9½-in-12, I fell back on my trusty Construction Master calculator (Calculated Industries, 22720 Savi Ranch Pkwy., Yorba Linda, Calif. 92687; 800-854-8075). The Construction Master "speaks" in rise-per-foot rather than in sine/cosine, so you don't have to know trigonometry to use it. I came up with an unadjusted rafter length of 10 ft. 2⁷⁄₁₆ in.

So at a point on the measuring line 10 ft. 2⁷⁄₁₆ in. from the corner of the bird's mouth, I drew a plumb line representing the unadjusted length of the common. To compensate for the 4½-in. thickness of the ridge, I drew another plumb line back from the unadjusted length by a 2¼-in. shortening allowance—half the ridge thickness.

As a result of the cross-gable configuration, only the first rafters in from each corner of the main roof were commons; the rest were jack rafters. These commons would anchor the lookouts for the gable overhang, though, so I doubled them.

Valley layout—Before laying out valleys, I prefer to rip the backing bevel on the upper edge of the valley stock (see sidebar, facing page). I usually bevel two pieces and nail them together later to make a double valley rafter with a V-trough down the middle. The backing bevel helps me orient the compound cheek cuts on both ends of the valley; cheek cuts go either outward or inward in relation to the center face of the valley.

If the ceiling below a valley is a cathedral-type, as was the case here, the lower edges of the valley stock should be beveled as well, with upper and lower edges parallel to each other. This keeps the underside of the valley rafter flush with the underside of the jack rafters and makes it easier to install the drywall. The vertical depth of the valley on both faces should be the same as the vertical depth of the commons and the jacks. (Vertical depth is the width of the rafter as measured along a plumb line.)

After ripping backing bevels on all the valley stock, I started laying out the first blind valley. I designated a top edge and a center face with a lumber crayon. At some arbitrary point on the center face, I drew a plumb line using the numbers 9½ and 17 on the square. I used 17 instead of 12 for the unit run because regular hips and valleys always run 17 in. diagonally for every 12 in. that the corresponding common runs perpendicular to the plate. The rise (in this case 9½) remains the same.

Measuring down from the top edge along the plumb line, I laid off the same raising distance I had found for the common rafters. Through the resulting point, I scribed a measuring line parallel to the rafter's edge. Starting at one end of the rafter, I laid off along the measuring line an allowance for the rafter tail. I had to leave more tail length for the valley rafter than for the common rafter because the valley tail, like the valley rafter, would have a greater run.

Once the tail allowance was established, I drew the bird's mouth with its corner on the measuring line, using 9½ and 17 on the square for plumb and seat cuts. From the corner of the bird's mouth, I laid off the unadjusted length of the blind valley. I got this number by using the HIP/VALLEY key on my Construction Master. With this key, I converted the length of the common rafter to the length of the valley rafter.

From the unadjusted length, I stepped back one half the thickness of the ridge measured at 45°. In this case, the diagonal thickness of the 4½-in. thick ridge was 6⅜ in., so I pulled the actual plumb cut back half that, or 3³⁄₁₆ in., from the unadjusted length. This adjustment, like all

Two methods of finding backing bevels

Beveling the top and bottom edges of a valley (or hip) rafter keeps them coplanar with the roof and the ceiling, which simplifies the installation of roof sheathing and drywall. I use two methods to find the backing bevels of hips and valleys: the scrap-block method and the graphic-development method.   — S. M.

Scrap-block method—To use the scrap-block method, begin by bisecting the angle formed by the adjoining walls where the rafter will sit. In the case of regular roof framing, that means bisecting a 90° corner at 45°. You can do this on the actual plates, but I usually just draw on a sheet of plywood or a piece of paper a 90° corner with a 45° line running through it.

Next, I cut a scrap block with the level seat cut of the valley at the lower end and a square cut at the upper end. The block doesn't have to be the same width as the actual valley, but it must be the same thickness. Set the block on the drawing, with its point on the vertex (photo below left). If the valley is a single 2x, the block should straddle the bisecting line, with its two faces offset ¾ in. to either side. If the valley is to be doubled, set one face of the block on the bisecting line.

From the point where the outside face of the block crosses the plate line, scribe a line on the face of the block parallel to the block's edge. This line indicates the downhill side of the backing bevel. The uphill side of the bevel will be either a center line drawn down the top edge of the block (in the case of a single 2x valley rafter) or the upper corner on the opposite face of the block (in the case of a double valley rafter). The angle is the same in either case.

On the end grain of the square cut, connect the downhill side of the backing to the uphill side (photo below right). This is the ripping angle for your circular saw.

Graphic-development method—The graphic-development method is the same process as the scrap-block method, but it's performed in two dimensions. Suppose the pitch of the valley is 9½-in-17. Starting with the same plan—a 90° angle bisected by a 45° line—apply a framing square with 17 at the vertex and 9½ on the bisecting line (drawing below). Scribe along the 17 side. This is the slope line and is essentially a view of the scrap block pushed over on its side. Next, draw a perpendicular line at any point along the slope line, until it hits the bisecting line. This is a side view of the square cut you made on the scrap block.

Where the perpendicular line hits the bisecting line—point X—extend lines perpendicular to the bisecting line in both directions until they hit the plate lines at points A and B. The distance from A to B is analogous to the thickness of the valley. With a compass at point X, swing the original perpendicular line down in an arc to hit the bisecting line and connect the resulting point C with A and B. Now imagine you're looking at the end grain of the square cut you made on the scrap block; lines BC and AC represent the lines you drew connecting the downhill side of the backing bevel to the uphill side. Angle ACX is the circular-saw tilt angle.

For an irregular plan, when the walls intersect at some angle other than 90°, the bisecting line will not be 45°. Otherwise, the procedure for finding the backing bevels is the same.

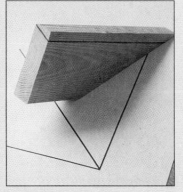

The scrap-block method. Beveling the edges of a valley rafter makes it easier to nail on plywood and drywall. Here, a scrap block cut with the seat cut of the roof pitch is used to find the correct angle.

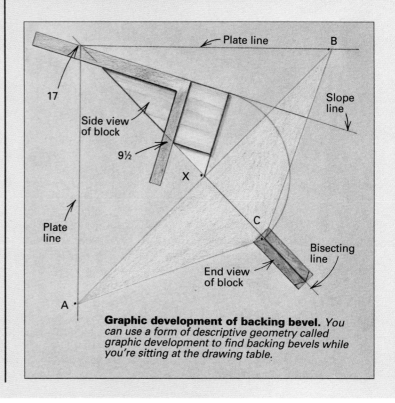

Graphic development of backing bevel. *You can use a form of descriptive geometry called graphic development to find backing bevels while you're sitting at the drawing table.*

---

shortening adjustments, was made in a horizontal direction, not along the measuring line.

I now had the valley rafter's true length, but I still needed to ascertain the direction of the bevels for the rafter's two compound plumb cuts—the first one located at the back of the bird's mouth where it would fit up against the edge of the plate, and the other at the top of the valley where it would bear against the ridge. After checking the plan, I looked down on the edge of the valley rafter and visualized its position in the completed frame. I then made crayon marks to indicate whether the bevels would go inward or outward from the center face. I cut one half of the double valley and used it as a template for the other half, being careful to orient the bevels

in their correct relationship to the center face; the two halves were opposite in this regard.

I cut another pair of rafter halves for the blind valley on the other side of the roof. This pair was the mirror image of the first, with the bevels going in the opposite direction. When both pairs of blind valleys were cut and nailed together, we hauled them up to the roof for the acid test. I got a lot of grief from the crew for all my ciphering, so I was relieved when both valley rafters dropped perfectly into place.

Short valleys, low ridges and jacks—The short valleys were laid out in much the same way as the blind valleys into which they would butt, though with a few differences. Their length was

extrapolated via calculator (the HIP/VALLEY key again) from the length of the lower-gable common rafter instead of from the upper-gable common rafter. This relationship is evident in the framing plan on p. 21.

You can also see from the plan that the short valley butts squarely into the blind valley, which seems peculiar if you're used to the oblique orientation of most valleys. Consequently, the plumb cut at the top of the rafter was made with the saw set square, as for a common, and the shortening allowance was half the thickness of the blind valley rafter measured at 90° (not the 45° thickness).

The inboard ends of the short lower ridges, where they nuzzle into the intersection of the

A slight adjustment. Where the blind valley extends above the lower ridge, the backing bevel had to be reversed on one side (the left side in the photo above) so that it wouldn't break the plane of the roof. The author scored this section of the rafter with a saw and used an ax to hew it flush with the roof.

A little insurance. To strengthen the connection between roof and walls, steel brackets were added between the top plates and the longest jack rafters, which were doubled up and were the only rafters with collar ties.

Cathedral ceiling. The careful framing of the roof makes possible the crisp lines of a cathedral ceiling. Exposed collar ties keep the walls from spreading. Photo by Kevin Ireton.

blind valley and the short valley, got a double 45° bevel cut made square across their faces. Once the short valley rafters and lower ridges were nailed in place, I had to make an adjustment to the blind valleys. Where the blind valleys extend above the lower ridge, the backing bevel had to be reversed on the side closest to the short valley. I scored it with a saw and used an ax to hew it flush with the main roof (top left photo, above).

A cross-gable roof is mostly jack rafters, or "fill" as they're referred to on the West Coast. Although methods exist for cutting sets of jacks in diminishing progression, I find that the accumulation of error produced by this system makes it more trouble than it's worth. I just lay off the positions of the jacks on the ridge and on the valley, then measure in between. With a lumber crayon, I scribble the measurements on the ridge beam large enough to read them from the ground.

If the addition's plan had been a square, the outward thrust of the valley at each corner would have been resisted by a pair of walls perpendicular to each other. If the walls were adequately tied together, neither collar ties nor structural

ridges would have been necessary. But the plan was rectangular, and we worried about the valleys pushing against the long walls a couple of feet in from the corners. A stronger ridge was one alternative, but strengthening the ridge would have been difficult without making it deeper and bringing it below the ceiling planes. So to tie the opposing long walls together, we bolted clear fir collar ties between the longest pair of jack rafters at each end (photo above right). We used steel angle brackets, cut from heavy angle stock, to strengthen the connection between the jacks and the walls (bottom left photo, above).

Vents and vergeboards—Venting a cross-gable roof that has a cathedral ceiling is problematic because there's little or no eaves soffit to provide cool-air intake. We had only one bay at each end vented at the eaves, but by taking a notch out of the top edge of each jack rafter toward its lower end, we managed to get at least a little draft in the bay's bordering valleys. I also could have recessed the top edge of the valley in relation to the jacks, as I sometimes do with hips, but this would have reduced its strength.

We vented the ridges with a concealed shingle-over type ridge vent. We vented the framed rake overhang by replacing one course of the yellow-pine wainscoting used as soffiting material with a strip of aluminum soffit vent.

The pierced and sawn vergeboards (see the photo on p. 20) are the dominant features of the exterior. We made the vergeboards from clear kiln-dried redwood 2x12 because any knots or checks would likely cause the delicate, short-grained pendants to break off. We laid out the vergeboard design using a single-repeat template traced from the existing house, adjusting the spacing to get an even number of pendants. Sawing them out was a chore, even with a heavy-duty jigsaw.

Instead of a finial, the vergeboards meet at a simple square shaft, turned catty-corner and suspended from the peak. I wanted to go wild with an ornate spire, but the architects held me back. Some things haven't changed in 140 years. □

Scott McBride is a builder in Sperryville, Va., and a contributing editor of Fine Homebuilding. *Photos by the author except where noted.*

Framing a Gable Roof

Cutting and stacking the way pieceworkers do it

by Larry Haun

The key to production roof framing is to minimize wasted motion. Here, carpenters nail gable studs plumb by eye—there's no need to lay out the top plates. The next step is to snap a line across the rafter tails and cut the tails off with a circular saw.

One of my earliest and fondest memories dates from the 1930s. I remember watching a carpenter laying out rafters, cutting them with a handsaw, and then over the next several days, artfully and precisely constructing a gable roof. His work had a fascinating, almost Zen-like quality to it. In a hundred imperceptible ways, the roof became an extension of the man.

But times change, and the roof that took that carpenter days to build now takes pieceworkers (craftspersons who get paid by the piece and not by the hour) a matter of hours. Since they first appeared on job sites, pieceworkers have given us new tools, ingenious new methods of construction and many efficient shortcuts. But what skilled pieceworkers haven't done is sacrifice sound construction principles for the sake of increased production. Quite the opposite is true; they've developed solid construction procedures that allow them to keep up with demand, yet still construct a well-built home.

The secret to successful piecework, from hanging doors (see *FHB* #53, pp. 38-42) to framing roofs, is to break down a process into a series of simple steps. To demonstrate just how easy roof framing can be (with a little practice), I'll describe how to cut and stack a gable roof the way pieceworkers do it.

The rafter horse—To begin with, pieceworkers try to avoid cutting one piece at a time. They'll build a pair of simple horses out of 2x stock so that they can stack the rafters on edge and mark and cut them all at once. To build the rafter horses, lay four 3-ft. long 2x6s flat and nail a pair of 2x blocks onto each, with a 1½ in. gap between them so that you can slip in a long 2x6 or 2x8 on edge (photo left, following page). An alternate method is to cut a notch 1½ in. wide by about 4 in. deep into four scraps of 4x12. Then you can slip a long 2x6 or 2x8 on edge into these notches. Either of these horses can

easily be broken down and carried from job to job. The horses hold the rafters off the ground, providing plenty of clearance for cutting.

Cutting the rafters—Rafters can be cut using a standard 7¼-in. sidewinder or worm-drive circular saw. This isn't the first choice for most pieceworkers, who prefer to use more specialized tools (especially when cutting simple gable roofs). But it is the more affordable choice for most custom-home builders. If you are using a standard circular saw, load rafter stock on the horses with their crowns, or convex edges, facing *up*—same as the rafters will be oriented in the roof frame. Determine which end of the stack will receive the plumb cuts for the ridge and flush this end. An easy way to do this is to hold a stud against the ends and pull all the rafters up against it. Then measure down from this end on the two outside rafters in the stack and make a mark

Pieceworkers typically build a pair of simple portable rafter horses (photo above). The horses allow them to stack rafters off the ground on edge so that they can mark and cut the rafters all at once.

One way to lay out rafters is with a site-built rafter tee (photo above). The fence on top of the tee allows easy scribing of the ridge cuts and birds' mouths. When laid out this way, rafters are cut one at a time with a circular saw.

corresponding to the heel cut of the bird's mouth (the notch in the rafter that fits over the top plate and consists of a plumb heel cut and a level seat cut). Snap a line across the tops of the rafters to connect the marks.

Next, make a rafter pattern, or layout tee, for scribing the ridge cuts and birds' mouths (right photo above). I usually start with a 2-ft. long 1x the same width as the rafters. Using a triangular square such as a Speed Square (The Swanson Tool Co., 1010 Lambrecht Rd., Frankfort, Il. 60423), scribe the ridge cut at one end of the template. Then move down the template about one foot and scribe the heel cut of the bird's mouth, transferring this line across the top edge of the template. This will serve as your registration mark when laying out the birds' mouths.

The layout of the bird's mouth on the tee depends on the size of the rafters. For 2x4 rafters, which are still used occasionally around here, measure 2½ in. down the plumb line and scribe the seat cut of the bird's mouth perpendicular to the plumb line. Leaving 2½ in. of stock above the plate ensures a strong rafter tail on 2x4 rafter stock. One drawback to this is that for roof pitches greater than 4-in-12, 2x4 rafters will have less than a 3½-in. long seat cut. Consequently, the rafters won't have full bearing on a 2x4 top plate. However, this presents no problems structurally as long as the rafters are stacked, nailed and blocked properly (the building code in Los Angeles requires a minimum bearing of 1½ in.). For 2x6 or larger rafters, you can make the seat cuts 3½ in. long without weakening the tails.

Once the layout tee is marked and cut, nail a 1x2 fence to the upper edge of the tee. This allows you to place the tee on a rafter and mark it quickly and accurately. Make sure you position the fence so that it won't keep you from seeing the ridge cut or your registration mark. Use the layout tee to mark the ridge cut and bird's mouth on each rafter. Scribe all the ridges first at the flush ends of the stock, sliding the rafters over one at a time. Then do the same for the seat cuts, aligning the registration mark on the template with the chalk marks on the rafters.

Next, with the rafters on edge, cut the ridges with your circular saw, again moving the rafters over one at a time. Then flip the rafters on their sides and make the seat cuts, overcutting just enough to remove the birds' mouths.

Production rafter-cutting—Cutting common rafters with production tools is both faster and easier than the method I've just described. In this case, you'll want to stack the rafters on edge, but with their crowns facing *down*. Flush up the rafters on one end and snap a chalkline across them about 3 in. down from the flush ends (the greater the roof pitch and rafter width, the greater this distance). The chalkline corresponds to the short point of the ridge plumb cut. Snap another line the appropriate distance (the common-rafter length) from this point to mark the heel cuts of the birds' mouths. Then measure back up from this mark about 2½ in. and snap a third line to mark the seat cut of the bird's mouth. This measurement will vary depending on the size of the rafters, the pitch of the roof and the cutting capacity of your saw (more on that later).

Now gang-cut the ridge cuts using a beam saw (top right photo, facing page). Blocks nailed to the top of the rafter horses will help hold the stack upright. My 16-in. Makita beam saw will cut through a 2x4 on edge at more than an 8-in-12 pitch and will saw most of the way through a 2x6 at a 4-in-12 pitch. To determine the angle at which to set your saw, use a calculator with a tangent key or, just as easy, look up the angle in your rafter-table book.

For steeper pitches or wider stock, make a single pass with the beam saw (or a standard circular saw) and then finish each cut with a standard circular saw, moving the rafters over one at a time. This way the only mark needed is the chalkline. The kerf from the first cut will accurately guide the second cut.

To make the process go even faster, apply paraffin to the sawblade and shoe. Also, try to stay close to your power source. If you have to roll out 100 ft. of cord or more, the saw will lose some power and won't operate at its maximum efficiency.

Another method for cutting ridges is to use the Linear Link model VCS-12 saw (Progressive Power Tools Corp., 303 N. Rose St., Suite 304, Kalamazoo, Mich. 49007). The model VCS-12 is a Skil worm-drive saw fitted with a bar and cutting chain that lets the saw cut to a depth of 12 in. at 90° (see *FHB* #39, p. 90). It's adjustable to cut angles up to 45° (top left photo, facing page). You can buy the saw or a conversion kit that will fit any Skil worm drive.

With the right tools, the birds' mouths can also be gang-cut with the rafters on edge. For the heel cuts, set your worm-drive saw to the same angle as the ridge cut and to the proper depth, and then make a single cut across all the rafters (bottom left photo, facing page). Seat cuts are made using a 7¼-in. or, better yet, 8¼-in. worm-drive saw fitted with a swing table. A swing table replaces the saw's standard saw base and allows the saw to be tilted to angles up to 68° (bottom right photo, facing page). I bought mine from Pairis Enterprises and Manufacturing (P. O. Box 436, Walnut, Calif. 91789). Set the swing table to 90° minus the plumb-cut angle (for example, 63½° for a 6-in-12 roof) and make the seat cuts, again in one pass.

The only drawback to using a swing table with a worm-drive saw is that it won't allow a substantial depth of cut at sharp angles, so it limits the amount of bearing that the rafters will have on the top plates (about 2½ in. maximum with an 8¼-in. saw). Again, this is of little concern if the roof is framed properly. Nevertheless, for jobs requiring a greater depth of cut, Pairis Enterprises just introduced a swing table to fit 16-in. Makita beam saws.

Gang-cutting birds' mouths works especially well because you needn't overcut the heel or the seat cut, which weakens the tail. Once you get used to working with these production tools, you'll find that it takes longer to stack the rafters than to cut them.

An even faster way to make the seat cuts is to use an 8¼-in. worm-drive saw equipped with a universal dado kit, a rig that has been around the tracts for over 15 years (middle photo, facing page). The dado kit (also manufactured by Pairis Enterprises) consists of an accessory arbor that fits on the saw, allowing it to accept a stack of carbide blades up to 3¼ in. thick. With this setup, birds' mouths can be gang-cut in a single pass and require just one chalkline for the heel cut.

The rig is surprisingly easy to control as long as it's used for its intended purpose, which is to plow out stock on a horizontal surface. In use, the rig whines like a router and hurls big chunks of stock out the front end. Though the guard effectively prevents wood chips from hitting the operator in the face, it's particularly important to wear safety glasses when operating this tool. The only drawback to this dado saw is its cost—about $750 including the saw. But if you cut a number of roofs a year, it will pay for itself in short order.

With the rafters cut, you can now carry them over to the house and lean them against the walls, ridge-end up. The rafter tails will be cut to length in place later.

Staging and layout—Now it's time to prepare a sturdy platform from which to frame the roof. The easiest way is to simply tack 1x6s or strips of plywood across the joists below the ridge line to create a catwalk (the joists are usually installed before the roof framing begins). Run this catwalk the full length of the building. If the ridge works out to be higher than about 6 ft., pieceworkers will usually frame and brace the bare bones of the roof off the catwalk and then install the rest of the rafters while walking the ridge.

For added convenience, most roof stackers install a hook on their worm-drive saws that allows them to hang their saw from a joist or rafter. When not in use, the hook folds back against the saw and out of the way (for more on these saw hooks, see *FHB* #55, p. 92).

The next step is to lay out the ridge. Most codes require the ridge to be one size larger than the rafters to ensure proper bearing (2x4 rafters require a 1x6 or 2x6 ridge). Make sure to use straight stock for the ridge. In the likely event that more than one ridge board is required to run the length of the building, cut the boards to length so that each joint falls in the center of a rafter pair. The rafters will then help to hold the ridge together. Let the last ridge board run long—it will be cut to length after the roof is assembled.

Be sure to align the layout of the ridge to that of the joists so that the rafters and joists will tie together at the plate line. If the rafters and joists are both spaced 16 in. o. c., each rafter will tie into a joist. If the joists are spaced 16 in. o. c. and the rafters 24 in. o. c., then a rafter will tie into every fourth joist. Regardless, no layout is necessary on the top plates for the rafters. Rafters will either fall next to a joist or be spaced the proper distance apart by frieze blocks installed between them. Once the ridge is marked and cut, lay the boards end to end on top of the catwalk.

Nailing it up—Installation of the roof can be accomplished easily by two carpenters. The first step is to pull up a rafter at the gable end.

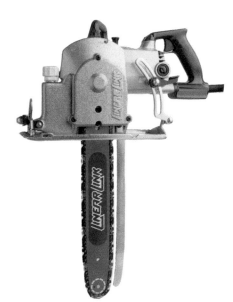

To save time, ridges can be gang-cut with a 16-in. beam saw (photo right). Though these saws won't cut all the way through anything wider than a 2x4 at a 4-in-12 pitch, where necessary each cut can be completed using a standard circular saw. For these finish cuts, the kerfs guide the saw. The only layout required is a single chalkline across the top edge of the rafters. An alternate method is to use a Linear Link saw (photo left), a Skil worm-drive saw fitted with a bar and chain.

Photo: Eric Haun

One method for gang-cutting birds' mouths is to cut the heels with a worm-drive saw (photo left) and the seats with a worm-drive saw fitted with a swing table, an accessory base that adjusts from -5° to 68° (photo right). By equipping a saw with a universal dado kit (photo above), birds' mouths up to 3¼ in. wide can be plowed out in a single pass.

While one carpenter holds up the rafter at the ridge, the other toenails the bottom end of the rafter to the plate with two 16d nails on one side and one 16d nail (or backnail) on the other. The process is repeated with the opposing rafter. The two rafters meet in the middle and hold each other up temporarily, unless, of course, you're framing in a Wyoming wind. If that's the case, nail a temporary 1x brace diagonally from the rafters to a joist.

Next, move to the opposite end of the first ridge section and toenail another rafter pair in the same way. Now reach down and pull up the ridge between the two rafter pairs. There is no need to predetermine the ridge height (photo facing page). Drive two nails straight through the ridge and into the end of the first rafter, then angle two more through the ridge into the opposing rafter. To keep from dulling a sawblade when you're sheathing the roof, avoid nailing into the top edge of a rafter. At this point, nail a 2x4 leg to the ridge board at both ends to give it extra support. If these legs need to be cut to two different lengths to fit beneath the ridge, it means that the walls probably aren't parallel and, consequently, that the ridge board isn't level. In this case yank the nails out of the rafter pair at the top plate on the high end of the ridge and slide out the rafters until the ridge is level. The key to avoiding all this hassle is, of course, to make sure the walls are framed accurately in the first place.

Next, plumb this ridge section. This can be accomplished in a couple of ways. One way is to nail a 2x4 upright to the gable end ahead of time so that it extends up to the height of the ridge. This allows you to push the end rafters against the upright and to install a 2x4 sway brace extending from the top plate to the ridge at a 45° angle. This is a permanent brace. Nail it in between the layout lines at the ridge.

A second method is to use your eye as a gauge. Sighting down from the end of the ridge, align the outboard face of the end rafters with the outside edges of the top and bottom plates, and then nail up a sway brace. Either way, the ridge can be plumbed without using a level. This means carrying one less tool up with you when you stack the roof.

With the bare bones of the first ridge section completed, raise the remaining ridge sections in the same way, installing the minimum number of rafter pairs and support legs to hold them in place. When you reach the opposite end of the building, eyeball the last rafter pair plumb, scribe the end cut on the ridge (if the ridge is to be cut at the plate line), slide the rafters over a bit and cut the ridge to length with a circular saw. Then reposition the rafters and nail them to the ridge. Install another sway brace to stabilize the entire structure.

Now stack the remaining rafters, installing the frieze blocks as you go. Nail through the sides of the rafters into the blocks, using two 16d nails for up to 2x12 stock and three 16d nails for wider stock. Where a rafter falls next to a joist, drive three 16d nails through the rafter into the joist. This forms a rigid triangle that helps to tie the roof system together.

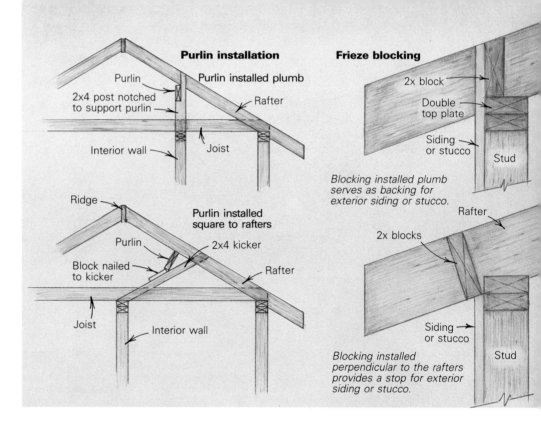

Purlin installation

Purlin — Purlin installed plumb
2x4 post notched to support purlin
Rafter
Interior wall — Joist

Ridge
Purlin installed square to rafters
Purlin
2x4 kicker
Block nailed to kicker
Rafter
Joist
Interior wall

Frieze blocking

2x block
Double top plate
Siding or stucco
Stud

Blocking installed plumb serves as backing for exterior siding or stucco.

Rafter
2x blocks
Siding or stucco
Stud

Blocking installed perpendicular to the rafters provides a stop for exterior siding or stucco.

Roof-framing tips

Check your blueprints for the roof pitch, lengths of overhangs, rafter spacing and size of the framing members. But don't rely on the blueprints to determine the span. Instead, measure the span at the top plates. Measure both ends of the building to make sure the walls are parallel; accurate wall framing is crucial to the success of production roof framing.

Once you've determined the length of the rafters, compensate for the thickness of the ridge by subtracting one half the ridge thickness from the length of the rafters. Though theoretically this reduction should be measured perpendicular to the ridge cut, in practice for roofs pitched 6-in-12 and under with 2x or smaller ridges, measuring along the edge of the rafters is close enough. For 2x ridge stock, that means subtracting ¾ in. from the rafter length. An alternative is to subtract the total thickness of the ridge from the span of the building before consulting your rafter book.

Once you've figured the common-rafter length, determine the number of common rafters you need. If the rafters are spaced 16 in. o. c., divide the length of the building in feet by four, multiply that figure by three and then add one more. That will give you the number of rafters on each side of the roof. If there are barge rafters, add four more rafters. If the rafters are spaced 24 in. o. c., simply take the length of the building in feet and add two, again adding four more to the total if barge rafters are called for. —*L. H.*

In some parts of the country, blocking is not installed between the rafters at the plate. But in many areas, building codes require blocks. I think they're important. They stabilize the rafters, provide perimeter nailing for roof sheathing and tie the whole roof system together. They also provide backing or act as a stop for siding or stucco. If necessary, they can easily be drilled and screened for attic vents.

There are two methods for blocking a gable roof (drawings above right). The first is to install the blocking plumb so that it lines up with the outside edge of the top plate, allowing the blocks to serve as backing for the exterior siding or stucco. This requires the blocking to be ripped narrower than the rafters. The other method is to install the blocking perpendicular to the rafters just outside the plate line. The blocking provides a stop for the siding or stucco, eliminating the need to fit either up between the rafters. Also, there's no need to rip the blocking with this method, which saves time. Either way, blocks are installed as the rafters are nailed up. Sometimes blocks need to be cut a bit short to fit right. Rafter thickness can vary from region to region (usually it's related to moisture content), so check your rafter stock carefully.

Collar ties and purlins—In some cases building codes require the use of collar ties to reinforce the roof structure or purlins to reduce the rafter span (drawings above left). Collar ties should be installed horizontally on the upper third of the rafter span. They're usually made of 1x4 or wider stock, placed every 4 ft. and secured with five 8d nails on each end so that they tie the opposing rafters together.

Purlins should be placed near the middle of the rafter span. They can be toenailed to the rafters either plumb or square. If there's an interior wall beneath the center of the rafter span, install the purlin plumb and directly over the wall. This makes it easy to support

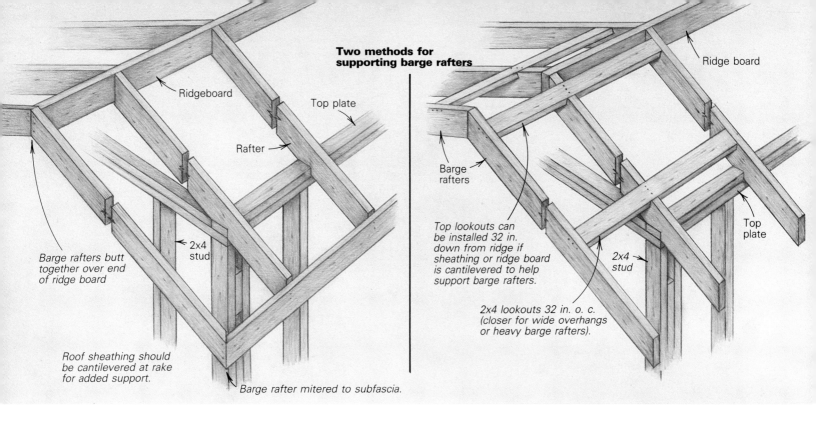

Two methods for supporting barge rafters

Ridgeboard

Top plate

Rafter

2x4 stud

Barge rafters butt together over end of ridge board

Roof sheathing should be cantilevered at rake for added support.

Barge rafter mitered to subfascia.

Ridge board

Barge rafters

Top lookouts can be installed 32 in. down from ridge if sheathing or ridge board is cantilevered to help support barge rafters.

Top plate

2x4 stud

2x4 lookouts 32 in. o. c. (closer for wide overhangs or heavy barge rafters).

the purlin with several 2x4 posts that bear on the top plate of the interior wall. The 2x4s are notched so that they both support the purlin and are nailed to the sides of the rafters.

If there isn't a wall beneath the center of the rafter span, toenail the purlin square to the rafters and install 2x4 kickers up from the nearest parallel wall at an angle not exceeding 45°. A block nailed to each kicker below the purlin will hold the purlin in place. Kickers are typically installed every 4 ft. Large purlins such as 2x12s require fewer kickers.

In some parts of the country, rafters have to be tied to the top plates or blocking with framing anchors or hurricane ties for added security against earthquakes or high winds. Check your local codes.

Framing the gable ends—Gable ends are filled in with gable studs spaced 16 in. o. c. Place the two center studs (on either side of the ridge) 14 in. apart. This leaves enough room for a gable vent, which allows air to circulate in the attic. Measure the lengths of these two studs, then calculate the *common difference* of the gable studs, or the difference in length between successive studs. Then you can quickly determine the lengths of the remaining studs. A pocket calculator makes it easy.

For a 4-in-12 roof pitch, the equation goes like this: $4 \div 12 \times 16 = 5.33$. Four equals the rise, 12 the run and 16 the on-center spacing. The answer to the problem, 5.33, or $5\frac{3}{8}$ in., is the common difference. Another way to calculate this is to divide the unit rise by three and add the answer and the unit rise together. For a 4-in-12 pitch, $4 \div 3 = 1.33 + 4 = 5.33$. For the angle cuts, set your saw to the same angle as that of the plumb cut on the rafters. Cut four gable studs at each length, and you'll have all the gable studs you'll need for both gable ends.

Once the gable studs are cut, nail them plumb using your eye as a gauge. There is no

need to lay out the top plates or to align the gable studs with the studs below. Be careful not to put a crown in the end rafters when you're nailing the gable studs in place.

Finishing the overhangs—The next step is to install the *barge rafters* if the plans call for them; these are rafters that hang outside the building and help support the rake. Sometimes barge rafters are supported by the ridge, fascia and roof sheathing. In this case, the ridge board extends beyond the building line so that the opposing barge rafters butt together over its end and are face nailed to it. At the bottoms the barge rafters are mitered to the sub fascia boards, which also extend beyond the building line. The roof sheathing cantilevers out and is nailed to the tops of the barge rafters.

Another way to support barge rafters is with lookouts. A lookout is a 2x4 laid flat that butts against the first inboard rafter, passes through a notch cut in the end rafter and cantilevers out to support the barge rafter (drawing above right). Lookouts are usually installed at the ridge, at the plate line and 32 in. o. c. in between (closer for wide overhangs or heavy barge rafters). If the roof sheathing cantilevers out over the eaves (adding extra support for the barge rafters), then the top lookouts can be placed 32 in. down from the ridge.

The notches in the rake rafters are most easily cut when you're working at the rafter horses. Pick out four straight rafters and lay out the notches while you're laying out for the birds' mouths and ridge cuts. Cut these notches by first making two square crosscuts with a circular saw 1½ in. deep across the top edges of the rafters. Then turn the rafters on their sides and plunge cut the bottom of the notch.

Lookouts are cut to length after they're nailed up. Snap a line and cut them off with a circular saw. That done, the barge rafters are face nailed

Pieceworkers don't waste time predetermining the ridge height. Instead, they toenail a pair of rafters to the top plates at either end of the ridgeboard, then raise the ridgeboard between the rafters and nail the rafters to it with 16d nails. A 2x4 sway brace is installed before the intermediate rafters are nailed up.

to the ends of the lookouts with 16d nails.

The final step in framing a gable roof is to snap a line across the rafter tails and cut them to length. Cutting the rafters in place ensures that the fascia will be straight. Use the layout tee or a bevel square to mark the plumb cut. If the rafters are cut square, use a triangular square. Then, while walking the plate or a temporary catwalk nailed to the rafter tails, lean over and cut off the tails with a circular saw. □

Larry Haun lives in Los Angeles and is a member of local 409, where he teaches carpentry in the apprenticeship program. Photos by Robert Wedemeyer except where noted.

Framing a Gable Roof Over a Bay

Letting ceiling joists cantilever beyond the walls makes this challenging roof detail stronger and easier to build

by Scott McBride

As an inveterate house watcher, I've always been curious about the roof-framing details on turn-of-the-century houses. Recently, I had a chance to examine one carpenter's techniques when the nearby Kilby farmhouse was being remodeled with an addition to match. The house was built in 1910 by a Virginia carpenter named John Mike Hawkins, who incorporated a dramatic two-story octagonal bay into each gable (top photo, facing page). These gable bays not only provide a focal point to the facade of the farmhouse but they also flood the rooms inside with daylight.

Because the bay spans the entire gable end of the house, the roof over the bay presents a unique roof-framing challenge. In order to maintain the regular gable roof without creating a hip where the bay walls angle inward, large triangular sections of soffit are left to extend out beyond the bay on each side. The roof above these soffit sections requires special attention because the rafters need to be suspended so far away from the supporting walls.

To solve this problem, John Mike installed a toe-board cornice (drawing right). This roof-framing detail was popular for many types of roofs, including the gable-over-bay. The technique emerged in the middle-Atlantic colonies in the 1600's and was used extensively into the early part of the 20th century. I'm surprised this method died out because it seems to offer some distinct advantages over the conventional bird's-mouth-on-plate system of framing a cornice.

Toe boards simplify cornice framing—In a toe-board cornice the ceiling joists overhang the outside walls of the house. The overhanging joist ends support the soffit, the fascia and the rafters. The toe board is a wide piece of framing lumber—in this case a 2x8—laid flat on the top outside ends of the ceiling joists. The toe board ties the joists together and provides a platform for the rafters to land on directly over the joists. Like the roof ridge, the toe board isn't being counted on to carry any weight; rather, it stabilizes the structure and aids in the construction process.

One of the chief advantages of the toe-board system is that no soffit lookouts are required. In conventional cornice framing, lookouts are the pieces of wood that run horizontally from the

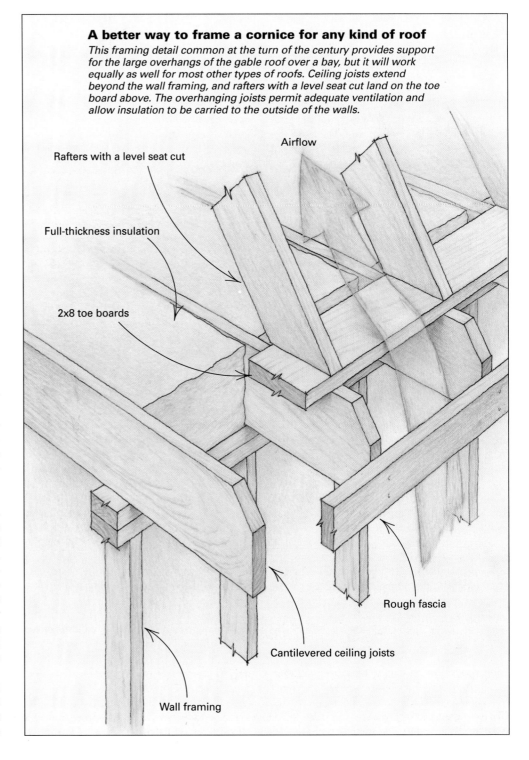

A better way to frame a cornice for any kind of roof

This framing detail common at the turn of the century provides support for the large overhangs of the gable roof over a bay, but it will work equally as well for most other types of roofs. Ceiling joists extend beyond the wall framing, and rafters with a level seat cut land on the toe board above. The overhanging joists permit adequate ventilation and allow insulation to be carried to the outside of the walls.

Rafters with a level seat cut

Airflow

Full-thickness insulation

2x8 toe boards

Rough fascia

Cantilevered ceiling joists

Wall framing

Drawing: Christopher Clapp

A dramatic facade, but a builder's challenge. The gable roofs on this turn-of-the-century farmhouse overhang the octagonal bays below, lending an impressive air to the facade. During construction of a recent addition, the builders copied the original cornice framing to support the cantilevered gable ends and rediscovered a useful technique that simplifies rafter and cornice construction while allowing full insulation over exterior walls.

Cantilevered ceiling joists extend beyond the top plates of the bay. As the walls of the bay angle inward, the overhang of the ceiling joists increases. The toe board on top of the joists ties everything together and provides solid nailing for the rafters.

A pleat in the roof connects two different pitches. The span of the gable roof over the bay was reduced to make the roof look more proportionate and less top heavy. To maintain the ridgeline, the builder created a jog in the eaves and a pleat in the roof that join the two different pitches. This pleat was framed with hip and valley rafters that run side by side from the corners of the jog and meet at the ridge (photo below).

The rafter ends get only one cut at the bottom. A single seat cut is all that is required for each rafter as it lands on the toe board above a corresponding ceiling joist.

Tapered blocking frames the pediment cornice. This attractive detail on the gable begins with tapered 2x blocks that are installed between the last ceiling joist and the rough fascia of the pediment. The blocking forms the pitched roof above the pediment cornice, and it provides sound nailing for the roof sheathing.

of the gable and the cornice fascia returning across its base. To support this overhanging pediment, the rough-fascia boards on both sides were extended past the last attic joist, and the skip sheathing on the roof was cantilevered to help support the barge rafter from above. Then tapered blocking was installed to span the gap between the rough fascia of the pediment and the last attic joist (photo left). This tapered blocking provides nailing for the soffit that's underneath, and it also forms the pitched roof above the pediment cornice.

A pleat in the roof improves proportions— John Mike Hawkins had executed the gable-over-bay detail on many houses, but something about it must have bothered him. Perhaps the large roof mass of the gable overhanging the slender column formed by the two-story bay looked somewhat precarious and gave the gable a top-heavy appearance. When he contracted to build the farmhouse for the Kilby family, Hawkins altered the design to improve its proportions.

His solution was to decrease the span of the overhanging gable slightly while maintaining the ridgeline of the main roof. This decision meant that the eaves' cornice would take a brief 45° jog around the ends of the bay (top photo, facing page). The break in the eaves combined with the continuous ridge required a subtle manipulation of the roof surface, sort of like a tailor sewing a pleat in a fine garment.

First, the ceiling joists and toe boards were stepped in to form the jog in the eaves. Then the bulk of the main roof-common rafters was set up with the end of the ridge cantilevered out over the bay. Framing the "pleat" on each side of the roof required an irregular hip and irregular valley placed side by side. (Don't worry: I'll spare you the mathematical machinations for figuring out the cheek cuts.) The bottoms of the valley and hip rafters strike the inside and outside corners of the jog in the eaves (bottom photo, facing page), and their tops converge at a point on the ridge.

Jack rafters fill in the roof surface between the valley rafter and the ridge and between the hip rafter and the eaves. The hip and valley rafters themselves are so close together that no jack rafters are required between them. The march of the common rafters resumes from the jog out to the overhanging gable end, but with one important modification. Because the span of the gable has been reduced and the height of the ridge is the same, the pitch of the common rafters above the bay is slightly steeper than the pitch of the main roof-common rafters (top photo, facing page). The final common rafter was doubled to accommodate the 2x4 framing of the gable.

John Mike's scaled-down gable produced its desired effect. Al Clarke and his crew duplicated Hawkins' original finish details on the farmhouse addition, including scrolled modillions and wagon-wheel gable louvers. In addition to being a home, the expanded Kilby house now serves as a day camp for kids. □

Scott McBride, a contributing editor of Fine Homebuilding, *lives in Castleton, Va., and is a Class-A building contractor. Photos by the author.*

rafter tails back to the walls to support the soffit. They are usually tedious to install and offer bouncy nailing at best. With narrow soffits I have sometimes spanned the short distance between the fascia and the nailer on the wall without using lookouts. But wider conventional cornices do require this additional framing to support the soffit material, especially when the soffit is split by a run of continuous soffit vent. Lookouts are unnecessary with a toe-board cornice because the ceiling joists provide solid nailing for the soffit material.

The toe-board cornice has other advantages as well. Because the rafter ends land on the toe board, they require only a single level seat cut, compared with the four cuts required for a conventional rafter tail (two for the bird's mouth and two for the tail). The wide toe board is safer to walk on while raising the rafters than the narrow wall plate. The toe board also is easier to get at with a hammer when nailing the rafters.

The rafter in a toe-board cornice is raised higher above the plate, leaving more room over the wall for insulation and creating better soffit-to-ridge airflow (drawing p. 30). In that respect the toe-board cornice acts like a raised-heel truss. The toe-board technique results in a higher soffit elevation that admits more light into the second-story windows, an advantage in cold climates but a possible liability in warmer ones. The high-

er soffit also allows more space for a wider frieze board between the soffit and the tops of the windows (top photo, p. 31). To my eye, a wide frieze board is attractive all by itself, but a wide frieze board also provides a nice backdrop for decorative corbels or brackets.

Cantilevered joists make roof framing independent of walls—Al Clarke, who remodeled the Kilby house, used John Mike's original work as a reference when he built the addition. Al began the roof framing of the addition by letting the ceiling joists overhang the second-story walls and fastening them together with the 2x8 toe boards. Each rafter landed on the toe board above its corresponding ceiling joist (bottom photo, facing page).

Where the gable roof extends over the bay, the toe boards prove particularly useful. As the walls of the bay angle inward, the wall plates that normally would carry the overhanging gable rafters get closer together. It would be difficult to frame a straight gable roof over the bay in the conventional manner. Instead, the overhang of the ceiling joists simply increases as the angled bay walls veer inward and the rafters land conveniently on top of the toe boards (bottom photo, p. 31).

John Mike created a pediment in the overhanging gable. This detail is recognized by the triangular shape formed by the rakes on the sides

Fine Gables

Ask your child to draw a house, and it's a good bet the result will feature a square box supporting a triangle roof. To most kids—and many adults—a gable is the very essence of "house." For all those who won't have a hip, here's a gaggle of gables in varying sizes and shapes.

Gables this page, clockwise from top: Aspen, Colorado; Kalamazoo, Michigan; Bethel, Connecticut; unknown.

Gables this page, clockwise from top: Bedford, Virginia; The Fortress of Louisbourg, Louisbourg, Nova Scotia; Newport, Rhode Island.

Framing Doghouse Dormers
Two ways to frame a basic gable dormer

by Scott McBride

I think of the dormer as one of the more playful aspects of a house. It wants to poke its head up and make a little mischief with the roof. The design of a dormer should echo the house's main roof in spirit, but not necessarily in detail. The Victorians enjoyed punctuating their roofs with all sorts of crazy outcroppings, and their adventurous spirit seems sadly missing from much of today's architecture.

The word "dormer" comes from the Latin verb *dormire*, to sleep. But while this suggests the dormer's function inside the house—to admit light and air into an attic bedchamber—it gives no indication of the aesthetic possibilities of this versatile architectural feature.

Dormers have been adapted for use on just about every style of house with a pitched roof. Although a few miss the mark, most succeed in lending some measure of character and charm to a home's appearance.

Gable dormers, because of their small scale, represent a microcosm of roof-framing theory and practice. And as such, they provide a good opportunity for novice carpenters and builders to study this complicated subject.

Walls—The basic gable dormer has a rectangular face wall and two triangular sidewalls, also known as "cheeks." The face wall is usually built up from the attic floor, or from an outboard header in the main roof, and the sidewalls are framed up from trimmer rafters in the main roof. (For more on dormer wall framing, see the article on pp. 42-46.)

As a general rule, the face wall of a gable dormer should be mostly window, with little or no siding on either side. A dormer should wink at you—not sit on the roof like a refrigerator with a mail slot. Bring the window rough opening right out to the corner posts. Since the gable end is non-bearing, you can usually omit the window jacks, and let the top plate define the height of the rough opening. Frequently, the corner board and exterior window casing are one piece, with solid trim covering the gable as well. This eliminates any siding on the face wall.

The triangle above a gable dormer window is traditionally a place where carpenters love to flaunt their woodworking skills. Appliqué, fancy-cut shingles and decorative truss work (stick style) are just a few of the treatments found here. Houses in the more formal Georgian and Adam styles often use the gable to field an elegant half-round fan window.

The exterior finish of the sidewalls can be

either siding material or roofing material, according to taste. Slate looks exceptionally good here, as do handsplit shakes.

Gable-roof styling—Getting the right pitch on the dormer roof is essential. You can do your planning on paper, but it's a good idea to mock up the roof lines after the walls are framed, either with 1x4s or by cutting out the dormer profile in a sheet of plywood. Since the dormer will usually be viewed from below (not straight on) elevation drawings are of limited value in judging its appearance. Whenever possible, use the main roof pitch for the dormer roof as well. This saves a lot of headaches in the framing.

Roof framing—The trickiest part of dormer construction is framing the intersection (valley) of the dormer roof and main roof. The type of ceiling inside the dormer determines which framing method to use. There are two choices.

The simpler approach, which I call the valley-board method, is to build the dormer roof on top of the main roof (drawing, facing page, left). Doing it this way means having a flat ceiling below the dormer because the main-roof cripple rafters cut off the dormer roof space from the rest of the attic.

I call the second approach the valley-rafter framing method. It has the inboard header set at

the elevation of the dormer ridge, rather than at the level of the wall plates (drawing, facing page, right). This allows for a cathedral ceiling inside the dormer. It is more trouble than the valley-board method, and requires a more thorough knowledge of roof-framing geometry.

Valley-board method—I'll begin with the simpler of the two approaches. After framing the dormer walls, set the inboard header (usually a pair of 2x6s or 2x8s) between the trimmer rafters with its bottom edge flush with the top of the dormer sidewall plates. This way the dormer ceiling joists will line up with the header.

Next, fill in the main-roof frame between the trimmer rafters by installing the cripple rafters, which extend from the inboard header up to the main roof ridge. These cripples have a plumb cut on both their upper and lower ends. If the cripples are long, it's a good idea to notch the lower ends so they hook over the header. This brings the weight of the cripple to bear on the top of the header. Otherwise, the strength of the joint depends solely on nails in shear. Metal framing connectors (sloped-seat joist hangers) can also be used to reinforce this connection.

If you sheathe the main roof before framing the dormer roof, you'll avoid the difficulty of cutting plywood around the dormer later. But you may want to leave the framing open to the dormer attic space in order to ventilate the roof bays and keep the insulation dry. In this case, go ahead and sheathe the roof, but don't nail off the plywood in the area of the dormer. After you've snapped the lines establishing the location of the valley, adjust the depth on your circular saw and cut out the plywood just inside the chalk lines.

The best way to lay out the dormer common rafters is to draw the elevation of the dormer gable full scale on the plywood subfloor (for more on rafter layout, see the articles on pp. 8-15 and pp. 72-77). Then you can establish the particulars of the cornice construction at the same time, and simply transfer the cutting angles to your rafter stock with a T-bevel.

Cut out four common rafters and use them to prop up a temporary ridge. With a straightedge, extend the line of the ridge over to the main roof. This will give you the location of the inboard end of the ridge and a point from which to measure the length of the ridge. If the ridge lands on a cripple rafter, its inboard cut will be at the same angle as the level cut of a main-roof common. If the dormer ridge falls between

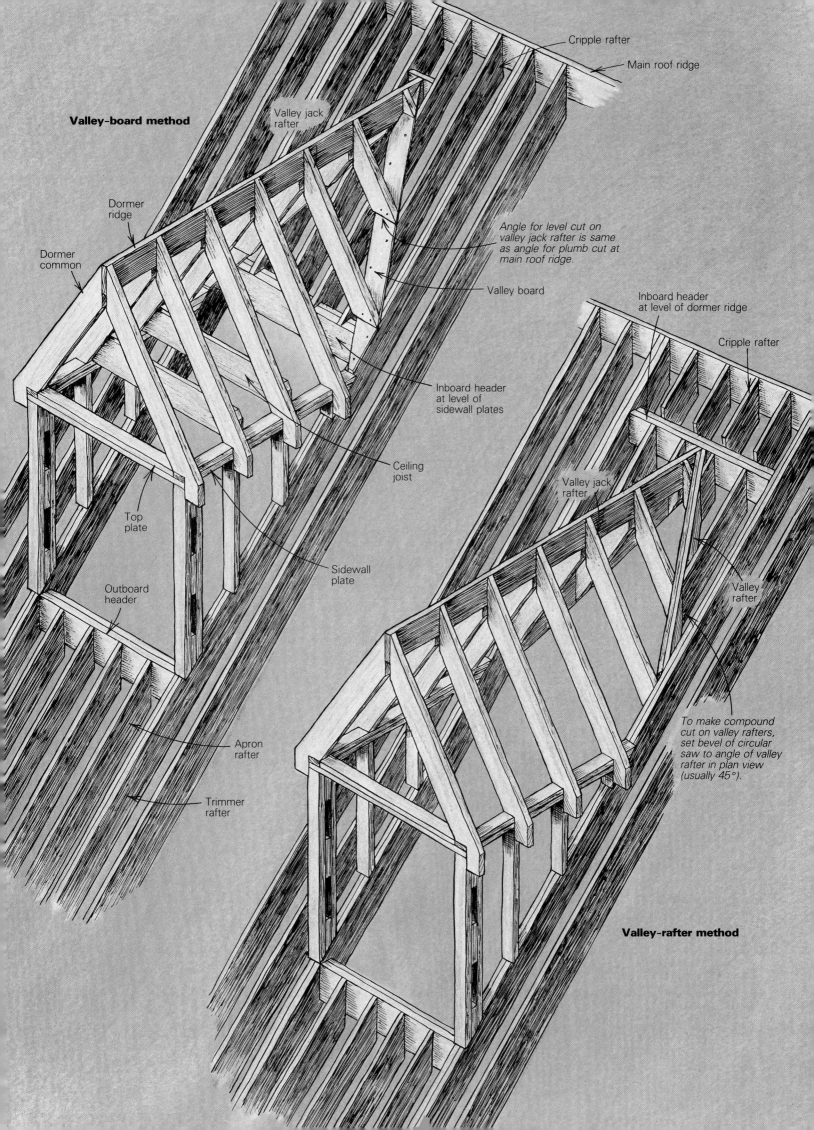

Valley-board method

Cripple rafter

Main roof ridge

Valley jack rafter

Dormer ridge

Dormer common

Angle for level cut on valley jack rafter is same as angle for plumb cut at main roof ridge.

Valley board

Inboard header at level of sidewall plates

Top plate

Ceiling joist

Outboard header

Sidewall plate

Apron rafter

Trimmer rafter

Inboard header at level of dormer ridge

Cripple rafter

Valley jack rafter

Valley rafter

To make compound cut on valley rafters, set bevel of circular saw to angle of valley rafter in plan view (usually 45°).

Valley-rafter method

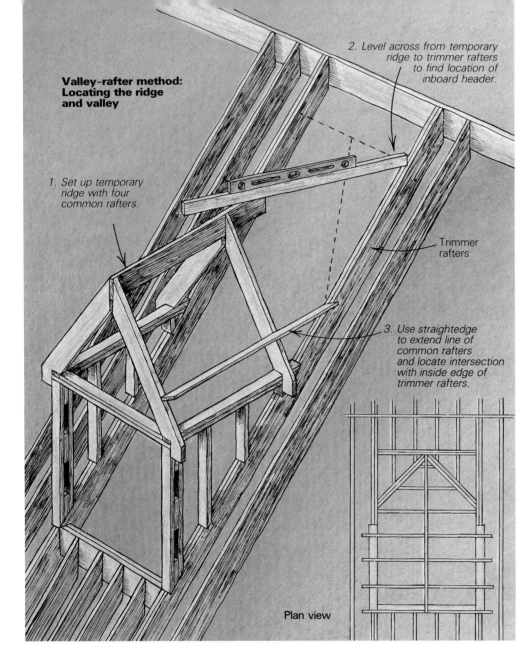

**Valley-rafter method:
Locating the ridge
and valley**

1. Set up temporary
ridge with four
common rafters.

2. Level across from temporary
ridge to trimmer rafters
to find location of
inboard header.

Trimmer
rafters

3. Use straightedge
to extend line of
common rafters
and locate intersection
with inside edge of
trimmer rafters.

Plan view

rafters, you will either have to install an extra cripple, or nail a block between two neighboring ones. In the latter case, the ridge would be cut square to fit against the block, as shown in the drawing at left on the previous page.

Cut the permanent ridge and nail it in place, along with the four common rafters, one pair at the gable end and the other pair at the inboard end of the plate. Leave the rest of the commons out until you've finished cutting and installing the valley jack rafters (the commons will just be in the way otherwise).

Placing a straightedge across the common rafters near the eaves, project the dormer roof plane onto the main roof frame or sheathing. Mark the point of intersection with a pencil. The exact location of this mark isn't critical, as long as it's in the dormer roof plane toward the bottom of the proposed valley. Snap a chalkline over the tops of the rafters from where the upper corner of the dormer ridge strikes the main roof down to your pencil mark.

If the layout of the dormer valley jacks is co-ordinated with the main-roof cripple rafters, each jack can bear directly over a cripple. In this case, the valley board can be 1x stock, or eliminated altogether. As long as you have direct

bearing on the cripple rafters, you can nail the jacks directly on the main-roof sheathing. If the bottoms of the valley jacks don't line up with the cripples, use 2x stock for the valley board to distribute the load. With a T-bevel, measure the angles for the ends of the valley board, which will be nailed to the main roof.

Since the valley board has thickness, it must be nailed back from the chalkline a bit so that its top outside corner will fall in the dormer roof plane. To determine the offset for the valley board, first take a scrap of 2x4 and put the jack-rafter seat cut on one end. To make this compound angle cut, lay out the level cut used for the dormer common on the face of the 2x4, and cut with the skillsaw set to the plumb-cut angle of the main-roof common rafter.

Tack the 2x4 to the valley board as if it were a valley jack and hold the valley board parallel to the chalkline on the main roof. Now extend a straightedge along the top of the 2x4 down to the chalkline. This will tell you how far to offset the valley board from the chalkline.

Lay out the jacks along the valley board by measuring 16-in. increments off the dormer common rafters closest to the main roof. As you measure, hold the tape or folding rule more or

less parallel to the dormer ridge, and perpendicular to the commons. Lay off the same spacing along the dormer ridge to correspond with the marks on the valley. Now you can directly measure the lengths of the jack rafters, from the uppermost point of the plumb cut on the top end to the toe of the seat cut on the downhill side. This is the longest dimension of the jack. Draw up a list of the lengths.

For each pair of jacks, you need to make only one compound-angle cut, or cheek cut; making the cheek cut on the end of one piece leaves the cheek cut for the opposing jack on the off-cut. Start with a board that is more than twice as long as the rafter you're cutting. After making the plumb cut on one end, measure off the rafter length and mark a level cut on the face of the rafter. Now tilt the circular saw to the plumb-cut angle of the main roof, and make the cut. It saves time to hook up two circular saws—one tilted for the cheek cuts, and one set square for the 90° cuts.

After installing the jacks, put in the remaining common rafters. Then install the ceiling joists and sheathing.

Valley-rafter method—Start by cutting out four common rafters and setting them up with a temporary ridge, just as you would for the valley board method. Extend a level line across from the ridge to the trimmer rafters (drawing, left). Nail the inboard header between the trimmers at this point, and install the main-roof cripple rafters above the header. Just be sure that their top edges are in line with the main-roof rafters.

Now measure for the dormer ridge, and cut and install it along with the four commons. Here, too, it's best to leave out the rest of the commons while you're working on the valleys.

The top of the valley is determined by the intersection of the dormer ridge and the main roof. It's a common mistake to think that the bottom of the valley will be where the sidewall plate strikes the trimmer. I've seen more than one textbook that shows it this way. In fact, that point will end up well below the surface of the dormer roof.

To locate the bottom of the valley, lay a straightedge across the two common rafters and project the line of the dormer roof onto the main roof. You have to find the point where the dormer roof plane intersects the inside edge of the trimmer rafter. This is where the centerline, along the top edge of the valley rafter, will meet the trimmer (drawing, left).

Although in terms of strength, dormer rafters usually need be no wider than 4 in., it's necessary to use stock the same width as the main-roof rafters if you want the dormer and main-roof ceilings to come together in a neat corner. This gives a cleaner look to the interior finish.

The rise of the valley rafter is the same as for the commons (think about it). But the run, on the other hand, is longer, just as it is for a hip rafter. If the pitch of the dormer roof is the same as the main roof (both are 7-in-12 in our drawing), and since they intersect each other at a right angle, the valley between them runs at a 45° angle to the common rafters in plan. This means the valley rafter runs 17 in. for every

12 in. of a common. Therefore, the pitch of the valley rafter is 7-in-17. Using these numbers (7 and 17) on a framing square, you can lay out the plumb-cut angles on the valley rafter. Then cut along the line where the 7 is.

Backing—Now we must determine the width of the valley rafter, and the bevel to be used for "backing" its top and bottom edges. Backing is the process of beveling a hip or valley in the same planes as the adjoining roofs, so that its thickness will not interfere with the sheathing. Hip rafters are usually dropped rather than backed. For the dormer valley rafter, you can skip the backing bevels on the top edge if you wish, by raising up the lower ends of the valley jacks a bit. This will cause the projection of their top edges to strike the centerline of the valley. If the dormer has a cathedral ceiling, however, backing is at least advisable on the bottom edge of the valley rafter. The attic and dormer ceilings will come to an outside corner here (photo next page), and the backing bevels will provide good seating for the drywall and sound nailing for the corner bead.

Backing hips and valleys also clarifies things when lining up the different members in the frame. Since a circular saw will handle the beveling without much trouble, I've found backing to be less bother than the guesswork involved in not backing, especially on tricky roofs.

To simplify the process, I often double the valley and hip rafters, even when doing so is not necessary for strength. This allows me to rip one bevel on each piece before assembly. When the two pieces are nailed up, the opposing bevels form either a concave V-trough or a convex ridge, which are the ideal forms for valley and hip rafters respectively.

Normally when you frame a roof you're concerned with lining up only the top surface of the framing members. The bottom doesn't matter because it's usually an unfinished attic. But when framing a dormer with a cathedral ceiling both the tops and bottoms of the rafters have to line up. The critical dimension is the vertical depth of the rafters measured along a plumb line. But since the valley rafter is rising at a shallower pitch than the commons, you need to cut it from wider stock than you used for your commons (drawing, top right).

To calculate the width you need for your valley rafter, first measure the vertical depth of a common rafter (the length of the plumb cut at the ridge is the easiest place to find this dimension). Then, using a framing square held at 7-on-17, draw lines that represent the pitch and plumb cut of the valley rafter on a sheet of plywood. Measure along the plumb-cut line the vertical depth of the common rafter and strike a line through this point, parallel to the pitch line. The distance between these lines is the width of your valley stock.

To calculate the backing angles for the valley rafter, start by laying the framing square on the face of the rafter stock, as in the drawing at middle right, with the tongue on 7 and the blade on 17 (the pitch of the valley). Now measure along the blade half the thickness of the valley stock, and mark this point on the rafter. Since I

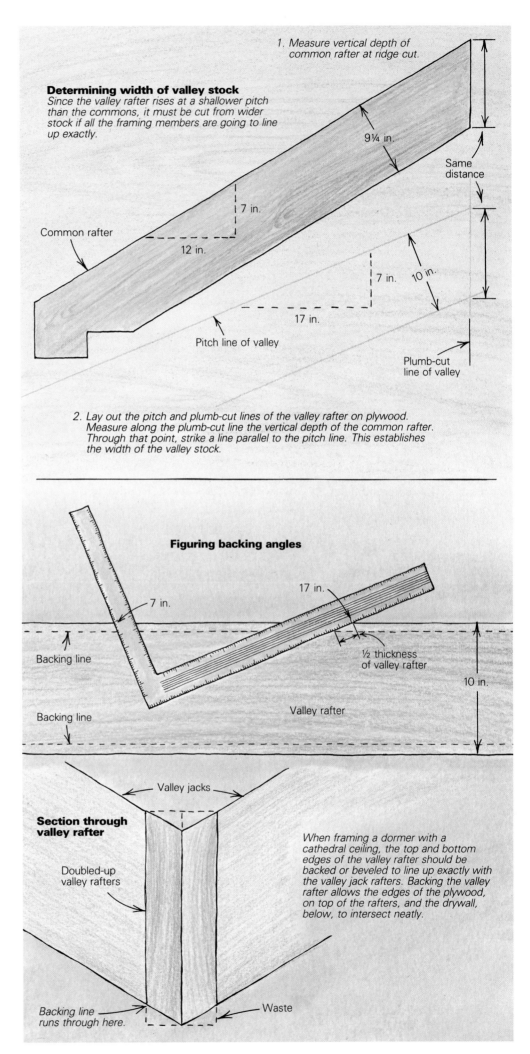

Determining width of valley stock
Since the valley rafter rises at a shallower pitch than the commons, it must be cut from wider stock if all the framing members are going to line up exactly.

1. Measure vertical depth of common rafter at ridge cut.

9¼ in.

7 in.
12 in.

Common rafter

Same distance

7 in.   10 in.

17 in.

Pitch line of valley

Plumb-cut line of valley

2. Lay out the pitch and plumb-cut lines of the valley rafter on plywood. Measure along the plumb-cut line the vertical depth of the common rafter. Through that point, strike a line parallel to the pitch line. This establishes the width of the valley stock.

Figuring backing angles

7 in.

17 in.

½ thickness of valley rafter

Backing line

10 in.

Backing line

Valley rafter

Section through valley rafter

Valley jacks

Doubled-up valley rafters

When framing a dormer with a cathedral ceiling, the top and bottom edges of the valley rafter should be backed or beveled to line up exactly with the valley jack rafters. Backing the valley rafter allows the edges of the plywood, on top of the rafters, and the drywall, below, to intersect neatly.

Backing line runs through here.

Waste

Although this dormer has a hip roof, it was framed with the valley-rafter method and illustrates the clean, geometric ceiling lines that make the complexity of this framing worthwhile.

use a separate 2x for each side of the valley, half the valley stock is 1½ in., but normally it's ¾ in. Strike a line through this point, parallel to the edge of the rafter. This is called the backing line, and the angle between it and the centerline on top of the valley rafter is the backing angle. (This rule holds only for "regular" hips and valleys, meaning those formed by roofs of equal pitch that join at right angles.)

Edge bevel and shortening adjustment—
As we have already seen, when equal-pitch roofs intersect at right angles, the resulting run of the valley (plan view) lies at 45° to both ridges (the main-roof header in this case is acting like a ridge). Therefore, making a cheek cut with the circular saw tilted to this angle will produce the correct bevel on the edge of the valley to fit the ridge.

Before it will fit though, the theoretical length of the valley must be shortened to allow for the thickness of the dormer ridge. Since the valley rafter is doubled, we will make the shortening adjustment on each half separately.

In order to fit against the ridge and inboard header, the valley rafter on the main-roof side of the valley gets two cheek cuts (see plan view, drawing at right, p. 38). First, make the cut that will fit against the header through the unadjusted length. No shortening adjustment is made here, because the theoretical layout line coincides with the face (not the centerline) of the header.

For the second cheek cut, measure back from the long point of the first cut, along a horizontal line, one-half the 45° thickness of the ridge (1¹⁄₁₆ in. for 1½-in. thick stock). Mark a plumb

line through this point and make the cut with the circular saw still set at 45°, beveling in the opposite direction to the first cut. If you wanted, you could make the double cheek cut here symmetrical by bringing the header out more to begin with. This would place the theoretical layout line down the middle of the outboard half of the header. But as a result, the header would have to be beveled where its corner protruded above the main-roof surface.

The ridge cut for the other half of the valley rafter simply gets shortened one-half the 45° thickness of the ridge. But if you put the first board in place, it will give you the point from which to measure the actual length of the second valley rafter board.

All these cuts are laid out using the same plumb-cut angle on the face. Shortening adjustments are always made horizontally—that is, perpendicular to the plumb cut.

Jacks—A gable dormer may be so small that it doesn't require any valley jack rafters. If it does require them, lay out the position of the jacks on the top edge of the valley rafter just as you would for the valley board. In this instance, however, the bottom end of the jack will be marked with a plumb cut on its face, instead of a level cut, and the circular saw will be set to 45°, instead of the main-roof plumb-cut angle. The top end of the jack will be cut the same as before—use the plumb cut of the dormer common, with the circular saw set square. □

Scott McBride is a carpenter and contractor in Irvington, N. Y.

Dormer crown molding

Crown molding is often used to trim gable dormers. Its curves and shadows add an elegant flourish to the dormer's overall composition. There are two basic applications. The first has one piece of crown molding along the eaves mitered to another along the rake, with very little overhang on either the eaves or the rake (figure 1, facing page).

The second application, which I call pedimented, continues the crown molding horizontally across the front of the gable, along with the rest of the cornice trim (frieze, bed, soffit and fascia). This forms the base of the pediment (figure 2). A sloped and flashed water table caps the cornice across the front to keep out rainwater. Two additional pieces of crown molding are run along the rake and die into the water table.

This type of pediment is traditional over entries and windows, as well as dormers, in the classically derived styles. Sometimes the cornice turns the corner and extends only a foot or so onto the gable end—just enough to provide a neat terminus for the rake crown. This is called a cornice return, or sometimes a boxed return. Pedimented gables are a lot of work, but their sharp appearance usually justifies the effort.

Turning the crown molding directly up the rake might seem the simpler of the two applications, but it isn't because of the miter. The crown molding changes planes as it turns upward, so if it's set in the usual way along the eaves, it won't match up with the molding along the rake. There are two ways to deal with this problem.

The first is simply to tilt the crown molding down at the eaves, so that its top edge lies in the roof plane, as shown in figure 3. Tapered blocking nailed to the fascia will give proper bearing for the back of the eaves-crown molding. It can then be mitered to the rake crown with a regular 45° miter.

You can get away with this approach on most low-pitched roofs. As the roof pitch of a dormer increases, however, this method forces the face of the crown molding to lie flat, contrary to its original intent.

The "correct" way to handle the situation is to have milled a molding with a slightly different profile just for the rake return, called a "raking molding" (not to be confused with the standard "rake molding" sold at lumberyards). This modified crown molding will have the same horizontal depth, front to back, as the crown molding at the eaves, but its vertical height will be stretched out a bit, depending on the pitch. The development of the rake crown's profile is shown in figure 4. The depth of the eaves crown at different points, front to back, is transferred to the rake crown unchanged, using arcs and parallel lines. Notice how the *diagonal* distances between points on the *face* of the eaves crown have become *vertical* distances between points on the *back* of the rake crown. When further projected onto the face of the rake crown, the diagonal distances between corresponding points increase. Note that the run of the eaves crown, shown in figure 4 as

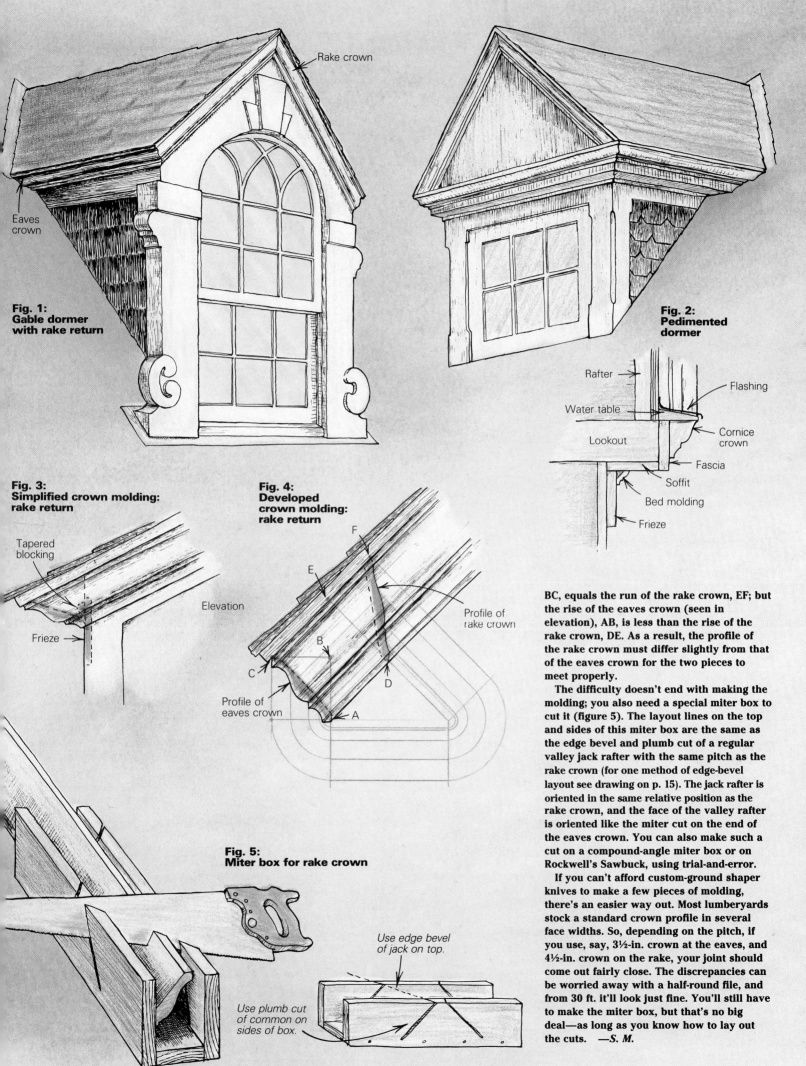

Rake crown

Eaves crown

Fig. 1:
Gable dormer
with rake return

Fig. 2:
Pedimented
dormer

Rafter

Flashing

Water table

Lookout

Cornice
crown

Fascia

Soffit

Bed molding

Frieze

Fig. 3:
Simplified crown molding:
rake return

Tapered
blocking

Elevation

Frieze

Fig. 4:
Developed
crown molding:
rake return

F

E

B

C

D

A

Profile of
rake crown

Profile of
eaves crown

BC, equals the run of the rake crown, EF; but
the rise of the eaves crown (seen in
elevation), AB, is less than the rise of the
rake crown, DE. As a result, the profile of
the rake crown must differ slightly from that
of the eaves crown for the two pieces to
meet properly.

The difficulty doesn't end with making the
molding; you also need a special miter box to
cut it (figure 5). The layout lines on the top
and sides of this miter box are the same as
the edge bevel and plumb cut of a regular
valley jack rafter with the same pitch as the
rake crown (for one method of edge-bevel
layout see drawing on p. 15). The jack rafter is
oriented in the same relative position as the
rake crown, and the face of the valley rafter
is oriented like the miter cut on the end of
the eaves crown. You can also make such a
cut on a compound-angle miter box or on
Rockwell's Sawbuck, using trial-and-error.

If you can't afford custom-ground shaper
knives to make a few pieces of molding,
there's an easier way out. Most lumberyards
stock a standard crown profile in several
face widths. So, depending on the pitch, if
you use, say, 3½-in. crown at the eaves, and
4½-in. crown on the rake, your joint should
come out fairly close. The discrepancies can
be worried away with a half-round file, and
from 30 ft. it'll look just fine. You'll still have
to make the miter box, but that's no big
deal—as long as you know how to lay out
the cuts. —*S. M.*

Fig. 5:
Miter box for rake crown

*Use edge bevel
of jack on top.*

*Use plumb cut
of common on
sides of box.*

Shed-Dormer Retrofit

Turning your attic into living space may be the remedy for your growing pains

by Scott McBride

Growing up amid the post-war baby plantations of central Long Island, I got to see a lot of expand-as-you-go housing. One of my earliest memories is the sight of slightly dangerous looking men, with hairy arms and sweaty faces, tearing the roof off of our home. My parents had decided to add onto our modest Cape, and that meant building a shed dormer. The following spring, a neighbor came over to take measurements; his house and ours, you see, were identical, and he wanted to do the same thing to his place. Before long, all the houses in our subdivision had sprouted the same 14-ft. long dormer.

Rivaled only by the finished basement, the enlarged and finished attic endures as the most practical way for the average suburban family to ease its growing pains. The shed dormer makes it possible to enlarge almost any attic space simply by flipping up the plane of the gable roof. Compared to the cost and complexity of a gable dormer, the shed dormer is a good choice where size and budget take precedence over looks.

Design—Shed dormers may be so narrow as to accommodate only a window or two, or may run the entire length of the house. In the latter case, it is common to leave a strip of the main roof alongside the rake at each gable end (photo above).

The trickiest part of designing a dormer is getting the profile right. To find the correct position of the inboard header and the dormer face wall, begin by making a scale drawing of the existing roof. Then draw in the dormer that you have in mind. What you're trying to determine here are the height of the dormer's face wall, the pitch of its roof, and where these two planes will intersect the plane of the main roof.

When determining the height of the face wall, consider exterior appearance, interior headroom, and window heights. The roof pitch you choose will affect the kind of roofing. Shingles require at least a 4-in-12 pitch. A flatter pitch should be roofed with 90-lb. roll roofing. This usually isn't a visual problem because you can't see the flatter roof from the ground.

Loading and bearing—Once you're sure that the existing ceiling joists will support live floor loads, you have to consider the other structural aspects of adding a shed dormer. Removing all or part of the rafters on one side of a gable roof upsets its structural equilibrium. You're taking a stable, triangulated structure and turning it into a not-so-stable trapezoid. The downward and outward forces exerted by the remaining rafters are no longer neatly countered by opposing members. The dormer's framing system has to compensate for this lost triangulation. To understand how this happens, let's take a look at a dormer's structural anatomy. As shown in the drawing on the facing page, the *inboard header* transfers loading from the *cripple rafters* out to the *trimmer rafters* on either side of the dormer. The full-length trimmer rafters send this lateral thrust down to the joists. With the main roof load reapportioned around the dormer, the new roof is structurally able to stand on its own.

On low-pitched dormers, the roof sheathing acts as a sort of horizontal beam that reinforces the inboard header and helps transfer the lateral thrust of the main roof out to the trimmer rafters. As you increase the pitch of the dormer, you decrease the ability of the dormer roof to act as a horizontal beam. And the lateral force of the dormer rafters themselves will sometimes threaten to bow out the dormer face wall. The solution is to tie the main-roof rafters and dormer rafters together with ceiling joists. These act as collar ties, creating a modified version of the original gable triangle.

We also have to consider vertical loads. The dormer roof on the outboard side is supported by the dormer face wall, which is built either directly atop the exterior wall or slightly to the inside, where it bears on the attic floor joists. This second option, shown in the drawing, facing page, leaves a small section of the original roof plane (called the apron) intact, and lets you retain the existing cornice and gutter. It also sets the dormer back a bit from the eave line and visually reduces the weight of the addition. Depending on the size of your floor joists, if the setback is more than one or two feet, the load on the attic floor joists can become too great. To lighten this load, you should install a header at the top of the apron to carry the roof load out to the trimmer rafters. In any case, this face wall will support a little more than half the weight of the dormer, depending on the roof pitch.

The other half of the dormer roof load usually rests on a large inboard header, which transmits the load through the trimmer rafters down to the exterior walls. To increase roof pitch and gain more headroom, the inboard header is frequently moved all the way up to the ridge of the main roof. If this ridge beam is made strong enough to carry roughly half the weight of the dormer

Retrofit framing details

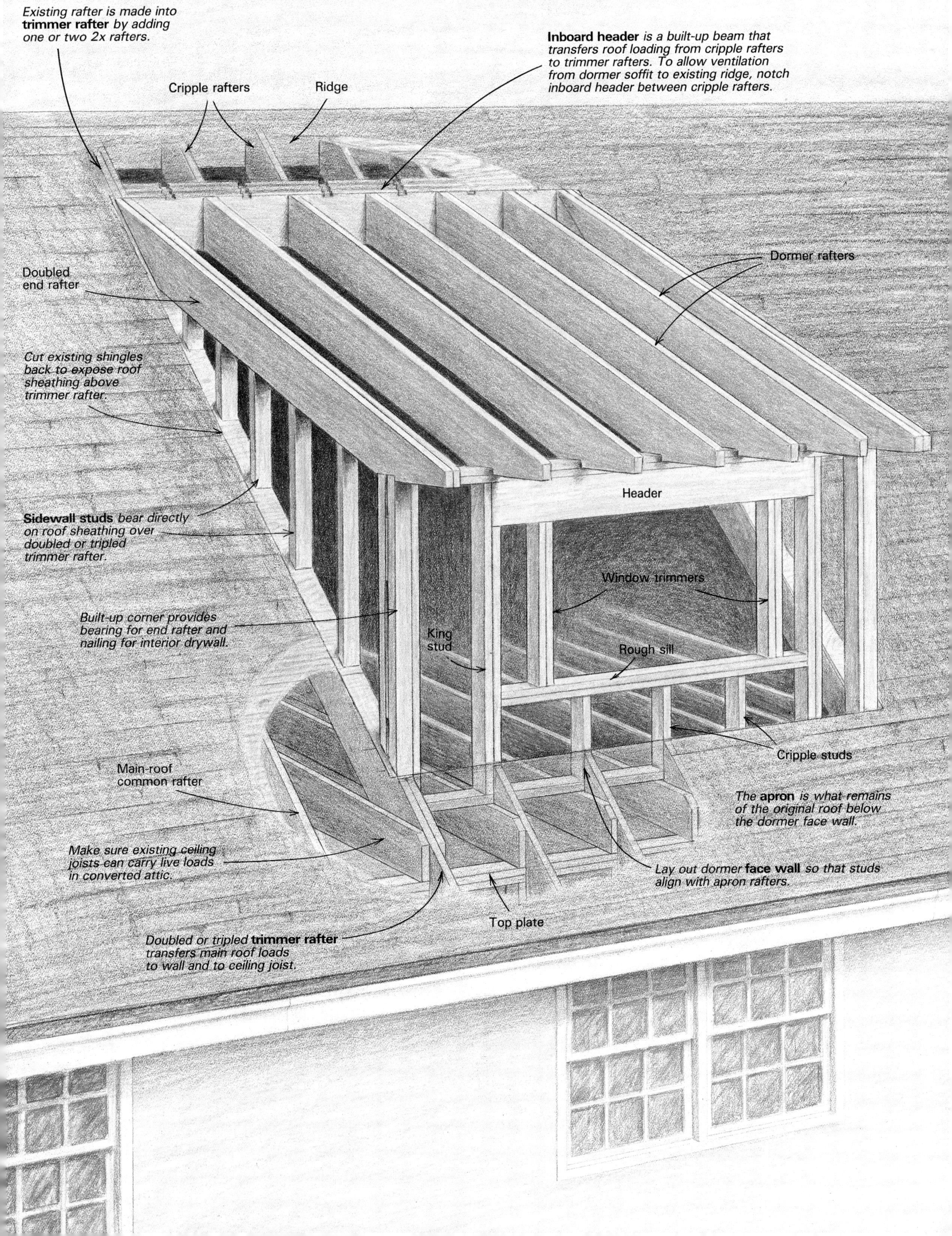

Existing rafter is made into **trimmer rafter** *by adding one or two 2x rafters.*

Cripple rafters

Ridge

Inboard header *is a built-up beam that transfers roof loading from cripple rafters to trimmer rafters. To allow ventilation from dormer soffit to existing ridge, notch inboard header between cripple rafters.*

Dormer rafters

Doubled end rafter

Cut existing shingles back to expose roof sheathing above trimmer rafter.

Header

Sidewall studs *bear directly on roof sheathing over doubled or tripled trimmer rafter.*

Window trimmers

Built-up corner provides bearing for end rafter and nailing for interior drywall.

King stud

Rough sill

Cripple studs

Main-roof common rafter

The **apron** *is what remains of the original roof below the dormer face wall.*

Make sure existing ceiling joists can carry live loads in converted attic.

Lay out dormer **face wall** *so that studs align with apron rafters.*

Top plate

Doubled or tripled **trimmer rafter** *transfers main roof loads to wall and to ceiling joist.*

roof and half the weight of the main roof, then it won't sag, and the rafters connected to it cannot spread apart at the plates. This allows the attic to have a cathedral ceiling.

If the ceiling is to be flat, the ceiling joists will prevent the roof from spreading, as mentioned earlier. In this case the ridge is non-structural and can be made of lighter stuff.

If you don't use ceiling joists and go for a cathedral ceiling, the length of your dormer will depend upon the practical length of the inboard header or structural ridge beam. About 12 ft. to 16 ft. is typical. At this length, a triple 2x10 or 2x12 should make an adequate header, capable of carrying half the dormer roof load, plus the weight of any cripple rafters above it. If the 2xs in the built-up header are slightly offset from one another and the main roof pitch is steep

enough, the header will not protrude below the ceiling. Sizes of all members should be checked by an engineer, architect or building inspector.

If you're going to build a long dormer, you can support the header between the trimmer rafters with an intermediate rafter. Hidden inside a partition wall that runs perpendicular to the face wall, this rafter picks up the load of the headers, which can then be reduced in size.

Preparation—Before you cut a big hole in your roof, you have to determine the location of the dormer from inside the attic. Lay some kind of temporary floor over the open joists to keep boots from going through ceilings and to keep trash out of the attic insulation.

You may want to use one of the existing rafters as a starting point and lay out the dormer from

there. In this case the existing rafter becomes a trimmer rafter and will have to be doubled or possibly tripled to carry the load. This can be done before the roof is opened up.

To lay out these extra trimmer rafters, measure the underside of an existing rafter from the heel of the plumb cut at the ridge down to the heel of the level cut at the plate. Transfer the respective angles with your T-bevel. These extra rafters don't support the cornice, so you don't have to cut a bird's mouth; just let the level cut run through. If a ceiling joist prevents the new rafters from reaching the plate, raise the level cut on the bottom of the rafters so they bear snugly on the top edge of the joist.

Now slide the additional rafters into the appropriate bays to make the trimmers. Any roofing nails protruding below the sheathing should

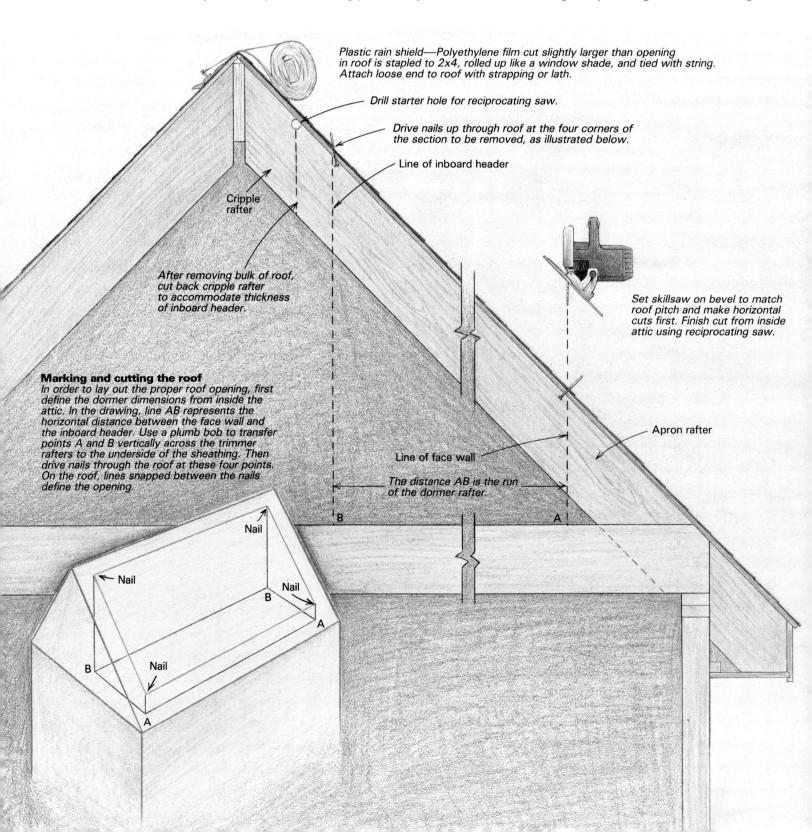

Plastic rain shield—Polyethylene film cut slightly larger than opening in roof is stapled to 2x4, rolled up like a window shade, and tied with string. Attach loose end to roof with strapping or lath.

Drill starter hole for reciprocating saw.

Drive nails up through roof at the four corners of the section to be removed, as illustrated below.

Line of inboard header

Cripple rafter

After removing bulk of roof, cut back cripple rafter to accommodate thickness of inboard header.

Set skillsaw on bevel to match roof pitch and make horizontal cuts first. Finish cut from inside attic using reciprocating saw.

Marking and cutting the roof
In order to lay out the proper roof opening, first define the dormer dimensions from inside the attic. In the drawing, line AB represents the horizontal distance between the face wall and the inboard header. Use a plumb bob to transfer points A and B vertically across the trimmer rafters to the underside of the sheathing. Then drive nails through the roof at these four points. On the roof, lines snapped between the nails define the opening.

Apron rafter

Line of face wall

The distance AB is the run of the dormer rafter.

B   A

Nail

Nail

Nail

B

Nail

B

Nail

A

A

be nipped off. Some persuasion may be necessary to bring the new rafter up tight against the old one. Spike the rafters together generously and toenail the new ones to the ridge and plate.

Now that you've defined the length of the dormer, you need to mark off the width. Begin by measuring from the outside of the exterior wall in to where you want the face wall (drawing, facing page). From here plumb a line up across the trimmer rafter and mark where this line intersects the roof. Now measure horizontally toward the ridge, from the proposed face wall to where you want the inboard header. Plumb another line up to the trimmer from here, and mark where it intersects the roof. This distance is the width of your dormer; it's also the run of the dormer rafter. Where these points (two on each set of trimmer rafters) touch the underside of the sheathing, drive four large nails up through the roof to mark the corners of the rectangular section you'll cut out from above. But before heading up to make the cuts, check for electrical wires, vent stacks and anything else you don't want your skillsaw to run into.

Rigging—Since houses with steep roof pitches make the best candidates for dormers, you'll need good rigging. Set up staging along the eaves, extending a few feet past both sides of the dormer location. If a hoist or pulley can be rigged in conjunction with the scaffold, so much the better. To gain access along the sides of the dormer, a ladder can be hooked over the ridge, or roof brackets can be set up.

"What happens if it rains?" is the question most often asked by clients. If proper precautions are not taken while the house undergoes dormer surgery, a heavy rainstorm could cause thousands of dollars in damage.

Once you get up to the ridge, install an emergency rain shield—a piece of heavy polyethylene film wrapped around a 2x4 somewhat longer than the length of the dormer. On the ground, spread out a piece of the poly several feet longer than the dormer and wide enough to reach from the ridge of the existing roof to the eaves. Staple one of the horizontal edges to the 2x4, and roll the sheet up like a window shade. Tie the roll with string, then carry it up to the roof and fasten its free edge to the ridge with wood lath or strapping. If it rains, cut the string and let the sheet unroll. The weight of the 2x4 hanging over the eaves will keep the poly tight, so it won't flap in the wind and so puddles won't develop.

Demolition—For cutting through the roof, you need a powerful skillsaw equipped with a nail-cutting blade. The heavy carbide tips on these blades are ground almost square, giving them the toughness needed to plow through asphalt, plywood and miscellaneous nails all at once. Some manufacturers coat this type of blade with Teflon to reduce friction. Eye protection is a must during this operation.

After snapping lines between the four nails you drove up through the roof, make the horizontal cuts first (there are only two of them and they're a little tougher to do). Set the skillsaw as deep as it will go, and set the saw's shoe to the plumb-cut angle of the main roof. Since you

cannot safely plunge-cut with a skillsaw when it's set on an angle, start the cut with your drill and reciprocating saw. Then use a slow, steady feed on the skillsaw. Keep moving, because the weight of the saw will tend to push the downhill side of the sawblade against the work, generating extra friction. Be particularly alert to the possibility of kickback; your blade will be crashing into 8d sheathing nails now and then. And remember, you're up on a roof.

To make the vertical cuts, set the skillsaw back to 90° and start the cut on the uphill end. These cuts are easier because the weight of the saw helps pull it through the cut. All you have to do is slide down the roof behind it. If you have had experience plunge-cutting, then begin this way. Otherwise, start the cut with the reciprocating saw, and finish with the skillsaw.

After the four outline cuts are done, make longitudinal cuts down the middle of each bay in the area to be removed. This divides the roof into manageable chunks. Before freeing these chunks by completing the cuts through the rafters, determine whether the remaining roof frame (the apron rafters below, and the cripple rafters above) need temporary support. If these pieces are short and well-nailed to the plate and ridge, they will stay up by themselves. If not, shore them up temporarily with 2x4 braces.

Now use your reciprocating saw inside the attic to complete the cuts through the rafters. Have a couple of burly helpers hold up each section while you're working on it. A 10-ft. 2x8 rafter, 14 sq. ft. of sheathing, and several layers of roofing make these chunks very heavy. The safest way to lower them to the ground is with a strong rope that's wrapped around a sturdy mast.

After removing the bulk of the roof, the bottom ends of the cripple rafters must be cut back to accommodate the thickness of the inboard header, without cutting through the sheathing. Drill a ¾-in. starter hole at the top of the mark with a right-angle drill. Then cut straight down with the reciprocating saw.

To finish up the demolition, you'll need to cut back the roofing material to make way for the dormer sidewalls. Set the depth of the skillsaw so that it will cut through all the roof shingles, but will just graze the sheathing. Snap longitudinal lines on the roofing, located back from the inside faces of the trimmers a distance equal to the width of the dormer sidewall framing plus sheathing thickness, plus ½ in. for clearance. Slice through the roof shingles along these lines, and peel back the roofing to expose strips of decking above the trimmer rafters. The dormer sidewalls will be built up from these, with the inside edges of the studs flush with the inside faces of the trimmer rafters.

Wall framing—Your next step will be to cut and lay out the plates for the face wall. In most situations, the bottom plate for the face wall will bear directly on the attic floor. You should lay out the face-wall framing so that the apron rafters will bear directly on the wall studs (aligning the framing in this way is called stacking).

In order to bring concentrated roof loads down safely onto the floor framing, window king studs also should be in line with the apron

rafters or else be located over a joist. You can then frame inward to get the necessary rough-opening width. Or you can forget all this and just double the bottom plate to distribute the load safely.

Taking the scaled measurements from your drawings, transfer the header and sill lengths onto the plates. The various stud lengths will also come from your drawings. Be sure to locate the rough sill for the windows at least several inches above the apron to keep rain and melted snow from creeping in underneath.

Cut all the face-wall components and assemble them on the attic floor. Then raise and plumb the wall, bracing it temporarily if necessary.

Next you have to fill out the corners of the face wall with a combination of beveled sidewall studs and blocking, as shown in the drawing on p. 43. The tops of these studs will be flush with the top of the face wall, and will give bearing to the dormer end rafters. Begin by cutting oversized pieces with the pitch of the main roof cut on one end. Stand these in place and mark their tops flush with the top of the face wall. Cut and nail. This completes the face wall.

Roof framing—The shed-dormer rafter is laid out just like any common rafter. The only differences are the generally lower pitch, and the fact that its plumb cut bears against a header instead of a ridge board. There are several ways to determine rafter length. I lay out the bird's mouth first, and then step off the rafter length with a large pair of dividers, using the method I described in my article on pp. 9-15. After marking the plumb cut at the ridge, lay out the rafter tail according to the soffit and fascia details from your elevation drawings. Then carefully cut out the rafter pattern. (For more on roof framing, and a glossary, see pgs. 8 and 140.)

Turning to the roof, first nail the inboard header to the trimmers. Joist hangers won't work in this situation because you'll want to offset the 2xs like stair steps, starting each one slightly above or below the next in order to fit the slope of the roof. Instead, just toenail each piece in place with plenty of 16d commons, and then spike them to each other.

Now try the rafter pattern at several different locations along the top plate of the face wall. If all is well, use the pattern to cut the rest of the rafters and nail them in place. The spacing of the rafters should align with the face-wall studs in the same way the studs align with the floor joists. The end rafters will have to be retrimmed on a sharp angle at the top, because they bear directly on the roof instead of on the header. Place one of the pattern-cut rafters in position, up against the trimmer rafter, and mark the roofline. After cutting, double the end rafters to provide nailing for drywall, or spike a 2x on the flat with its bottom face flush with the bottom edge of the end rafter.

The dormer sidewall studs are framed directly from the main roof up to the dormer's end rafters, without any plates. As with the corners of the face wall, begin by cutting oversized pieces with the pitch of the main roof cut on one end. Then stand the pieces in place and mark where they meet the dormer end rafter,

Piggyback shed dormers. **Shed dormers are often part of the original design on houses that have a gambrel roof. Here a second dormer was added on top of the first, probably to let more light into the room. The piggyback dormer also accommodates an air conditioner.**

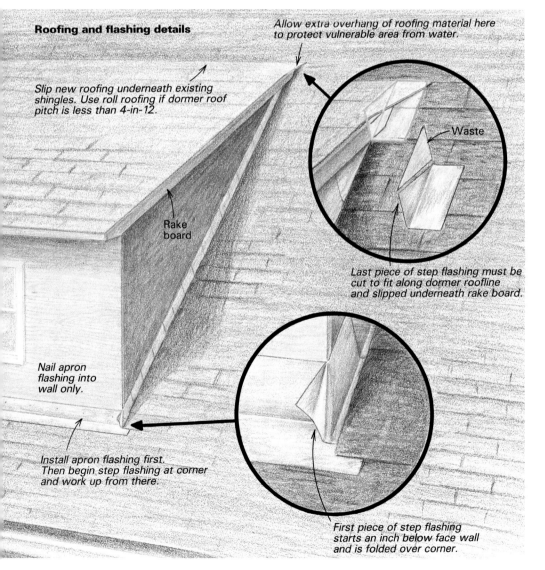

Roofing and flashing details

Allow extra overhang of roofing material here to protect vulnerable area from water.

Slip new roofing underneath existing shingles. Use roll roofing if dormer roof pitch is less than 4-in-12.

Waste

Rake board

Last piece of step flashing must be cut to fit along dormer roofline and slipped underneath rake board.

Nail apron flashing into wall only.

Install apron flashing first. Then begin step flashing at corner and work up from there.

First piece of step flashing starts an inch below face wall and is folded over corner.

cut along the mark and nail them up. These studs diminish in a regular progression, like gable studs, as they approach the ridge. If you don't want to mark each one in place, just mark the first two and measure the difference between them. This measurement is their common difference, and you can use it to calculate the diminishing lengths of the remaining studs.

Closing up—A few points on exterior finish are worth mentioning. Before decking the dormer roof, use a shingle ripper to remove nails in the first course of roof shingles above the dormer. This allows the dormer roofing material to be slipped underneath the existing shingles. If you install the sheathing first, the lower pitch of the dormer roof will interfere with the handle of the shingle ripper.

Flashing a shed dormer is relatively simple. As shown in the drawing below left, the apron is flashed first, then the dormer sidewalls are step-flashed. Use a 6-in. wide length of flashing along the apron, creased in half so that 3 in. of flashing runs up the dormer face wall and 3 in. extends over the apron shingles. Nail the face-wall side only. At the corners of the face wall, let the apron flashing run a few inches past the dormer sidewall. Slit the flashing vertically along the corner of the dormer, and push the overhanging vertical fin down flat on the main roof.

Overlap the apron flashing with the first piece of step flashing, where the dormer sidewall meets the main roof. Extend the step flashing down at least an inch past the face wall, and fold the vertical fin down and back on an angle. This will carry rainwater safely past the corner. You'll have to relieve the back of the corner board to fit over this first piece of step flashing.

Continue the step flashing all the way up the sidewall, slipping one piece of bent step flashing under the end of each roof shingle course, and pressing the other side up against the wall sheathing. Don't nail the step flashing into the roof; nail it to the sidewall only.

The rake board is usually furred out with a piece of 5/4 spruce so the siding can be slipped underneath. Where the dormer rake board dies into the main roof, the uppermost piece of step flashing is trimmed on an angle so that it can fit up behind the rake board and tight against the furring. Give the dormer roofing a little extra overhang here to help divert water from this sensitive spot.

Vents in the dormer soffit are a good idea. They prevent condensation in the dormer roof insulation as well as ice damming at the eaves. Since the inboard header blocks the flow of warm air at the tops of the roof bays, cut some notches across the top of the header in each bay or recess the top edge of the header slightly below the tops of the dormer rafters. If you're insulating between the rafters (instead of the ceiling joists), you'll need some spacers to create an airflow channel between the roof sheathing and the insulation. This allows some air flow from the dormer soffit vent to the ridge vent or gable-end louvers. □

Scott McBride is a carpenter and contractor in Irvington, N. Y.

Raising an Eyebrow

Two methods used to frame wave-like dormers

by James Docker

Eyebrow dormers had their American heyday during the late 19th century, when they turned up on the elaborate roofs of Shingle-style Victorian and Richardsonian Romanesque houses. Tucked between the conical towers, spire-like chimneys and abundant gables that distinguish these buildings, the little eyebrows provided a secondary level of detail to the roof and some much needed daylight to upstairs rooms and attics.

The roof cutters of that era could probably lay out an eyebrow dormer during a coffee break, but for a contemporary West Coast carpenter such as myself (well-versed in shear walls, production framing and remodeling techniques), framing an eyebrow dormer presented an out-of-the-ordinary challenge.

The setting for this dormer project was a rambling Tudor house in Atherton, California. The owners were adding a garage and remodeling several portions of the building, including a dilapidated barnlike recreation room next to the swimming pool. The roof of the house was covered with cedar shingles, and at the eaves and gable ends, curved shingles gave the roof a thatched look. Eyebrow dormers, rising by way of gentle curves from the plane of the 8-in-12 roofs, would look right at home on the house (photos right).

My job was to install five of them in the garage roof and a single larger eyebrow dormer in the roof of the recreation room. The garage ceiling would remain unfinished, so I didn't need to worry about providing backing for drywall or plaster. I would, however, have to solve that problem in the recreation room. The garage dormers required a lower level of finish while presenting the same conceptual problems, so I decided to build them first.

Rafter-type eyebrow—By the time I got on the job, contractor Dave Tsukushi had already taken delivery of the windows for the garage roof. They were arched, single-glazed units available off-the-shelf from Pozzi Wood Windows (Bend Millworks Systems, P. O. Box

Undulating courses of cedar shingles wrap over the warped contours of these eyebrow dormers on a Tudor-style house in Atherton, California. The solo dormer (top photo) lets light into the recreation room, while the others illuminate the garage (bottom photo).

Photos this page: Charles Miller

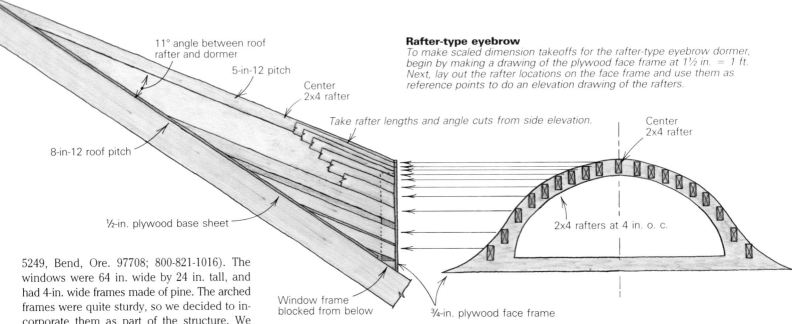

11° angle between roof rafter and dormer

5-in-12 pitch

Center 2x4 rafter

8-in-12 roof pitch

½-in. plywood base sheet

Window frame blocked from below

Rafter-type eyebrow
To make scaled dimension takeoffs for the rafter-type eyebrow dormer, begin by making a drawing of the plywood face frame at 1½ in. = 1 ft. Next, lay out the rafter locations on the face frame and use them as reference points to do an elevation drawing of the rafters.

Take rafter lengths and angle cuts from side elevation.

Center 2x4 rafter

2x4 rafters at 4 in. o. c.

¾-in. plywood face frame

5249, Bend, Ore. 97708; 800-821-1016). The windows were 64 in. wide by 24 in. tall, and had 4-in. wide frames made of pine. The arched frames were quite sturdy, so we decided to incorporate them as part of the structure. We faced them with ¾-in. ACX plywood, which would serve as a vertical surface for attaching the rafters, as well as backing for a stucco finish (top photo, right). At the top of the arch, this plywood face frame is 3¾ in. wider than the window frame. This dimension accommodates the 22½° plumb cut of a 2x4 rafter on a 5-in-12 slope—the pitch of our dormer. I screwed the face frame to the first window; then I braced it firmly on the roof, exactly on its layout between the roof trusses.

An arched window 2 ft. high with a base about 5 ft. wide makes a pretty tight curve for the dormer roof sheathing to follow. To make sure the curves stayed smooth and to ensure plenty of backing for the plywood, I decided to put my 2x4 rafters on 4-in. centers. I laid out their centerlines on the base of the window, and then used a level held plumb to transfer them to the arched portion of the face frame. Next I got out the string.

The luxury of full scale—Normally I make detailed drawings of unusual framing assemblies to familiarize myself with the geometry involved while still sitting on terra firma. This first dormer proved to be an exception to that rule, as I had the luxury of mocking it up on the garage roof. Still, clambering around on an 8-in-12 roof deck isn't everybody's idea of fun, so I would recommend doing a detailed drawing of the dormer's essential components, and then using it to scale the lengths and angles (more on this in a minute).

The opening in the roof made by an eyebrow dormer is bell-shaped in plan (bottom photo, right), with the bottom of the bell corresponding to the base of the window. Finding the shape of the bell became the next task.

First I tacked a couple of sheets of ½-in. ACX plywood to the roof deck (one over the other) so that their right edges were aligned with the centerline of the dormer. I specified plywood with an A side for all the plywood parts of the dormers (except for the sheathing) because it's much easier to draw accurate layout lines on a smooth, knot-free expanse of plywood than it is on a bumpy C or D side.

Then I cut a 2-ft. length of 2x4 with a 5-in-12

Plywood face frames screwed to arched window frames support the ends of the 2x4 rafters. The rafters are on 4-in. centers, and are parallel to one another. Note the uphill ends of the rafters. Their outside and inside corners touch the base and determine its shape.

Before the window and the rafters are installed, a bell-shaped base sheet is affixed to the roof deck. The next step will be to cut away the decking inside the base.

Drawings: Christopher Clapp

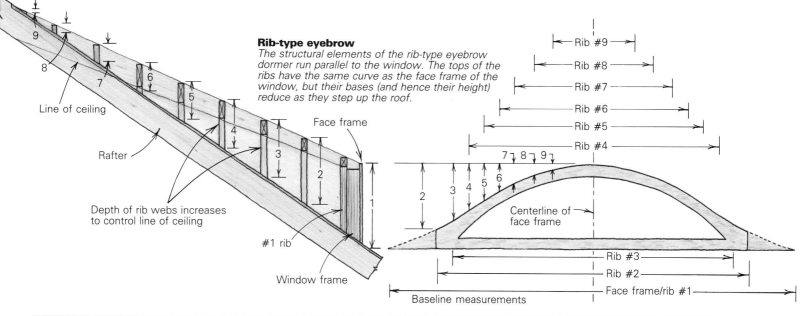

Rib-type eyebrow
The structural elements of the rib-type eyebrow dormer run parallel to the window. The tops of the ribs have the same curve as the face frame of the window, but their bases (and hence their height) reduce as they step up the roof.

Line of ceiling

Rafter

Depth of rib webs increases to control line of ceiling

#1 rib

Window frame

Face frame

Rib #9
Rib #8
Rib #7
Rib #6
Rib #5
Rib #4

Centerline of face frame

Rib #3
Rib #2
Face frame/rib #1

Baseline measurements

Rather than adjust the ribs on the roof, the author brought the roof to the workspace. Temporarily tacked on this 8-in-12 worktable, the corners of the rib bases are noted on the plywood to generate the bell-shaped base. The paper pattern will be used to cut out the plywood sheathing.

plumb cut on one end to act as a dummy rafter and rested the cut end on the top of the window frame. I held a string flush to the top edge of the dummy rafter, using a helper to hold it at the uphill end. This stringline represented the top of the center rafter, and its intersection at the roof was marked. Then the stringline-and-mark process was repeated for each rafter positioned to the left of center. By connecting the marks with a smooth curve, I had the outside line of the bell. To find the inside line of the bell, I measured the angle between the string and the roof

deck. This angle (11°) represented the cut needed on the bottom of the dormer rafters where they intersect the garage roof. Because all the rafters are at the same pitch and parallel to one another, this cut is the same for all the rafters. By making a sample cut on a short length of 2x4, I simply placed the tapered end on each rafter layout line and marked the inside corners. Connecting the dots gave me the inside of the bell curve. Because the rafter layouts are symmetrical, the bell-shaped base for half of the dormer is the mirror image of the other half.

Therefore, half of one base is all the layout template needed for one dormer. The same applies to the rafters. Once I had them measured and cut for half of the first dormer, I had templates for all the rest, allowing me to cut the parts quickly for all five dormers.

All this string-holding worked okay. But if I were to do this again, I'd do a drawing at a scale of 1½ in. to 1 ft. showing the elevation of the window from two vantage points (see drawing facing page). At this scale, it would be easy to make dimension and angle takeoffs for the lengths of the rafters and the angle at which they intersect the roof and the face frame. Then I'd use the rafters instead of the stringlines to figure out the bell shape of the plywood plate.

Assembly and sheathing—Once the plywood plates were cut out, we screwed them to the roof deck with galvanized drywall screws. Then we cut out the decking on the interior side of the base sheet, and screwed the eyebrow rafters to the plate and the face frame.

When one of the dormers had all its rafters, we draped 30-lb. felt over half of it and trimmed the felt along the valley formed by the intersection of the garage roof and the side of the dormer. This gave us the pattern we needed for marking cut lines on the ⅜-in. CDX plywood sheathing.

Bending a sheet of plywood over a radius this tight while holding it on the layout can be daunting—especially on a steep roof—so I wanted to prebend the largest pieces. Shallow kerfs on the underside of the plywood would have allowed it to bend more easily, but the structure of the dormers is visible from below so I wanted to avoid kerfs. Instead, I made a simple bending form out of a sheet of plywood with some 2x4 cleats nailed to the long edges. Then I stuffed several of ⅜-in. plywood sheets between the cleats, soaking each one liberally with the garden hose. Left in the form for a couple of days, the sheets took on a distinct curve, making them easier to bend over the rafters. Each dormer has one layer of ⅜-in. plywood affixed to the framing with 1-in. staples. We left the tops of the rafters unbeveled, but added beveled strips for better bearing where the sheets abut one another.

The garage roof had skip sheathing atop its decking to give the cedar shingles some breathing space. We carried the skip sheathing over each dormer by stapling a double layer of 6-in. wide redwood benderboard (5⁄16 in. thick) on top of the plywood. I had wondered what kind of valley flashing would be needed at the junction of dormer and roof. As it turned out, we didn't need any. The roofers wove layers of shingles together with very little exposure to form the valleys (bottom photo, p. 47).

After the rafters, the ribs—Unlike the multiple eyebrows on the garage roof, the single eyebrow atop the pool house had to have a finished ceiling underneath it. I decided that this extra wrinkle warranted another approach to the eyebrow's structure. Granted, you could hang blocking and furring strips from the bottom of a rafter-framed eyebrow to make a smooth transition from a flat ceiling plane to one with an arch, but why not make the bottom of the eyebrow structure conform as closely as possible to the shape of the arched portion of the ceiling? To that end, I worked up a full-scale drawing of the dormer (drawing previous page) on the recreation-room floor.

The window hadn't yet been ordered for this eyebrow, allowing me to design the arch from scratch. I made it long and low, taking the bulk of its face frame from a 10-ft. sheet of ¾-

The bell-shaped base of the rib dormer rests on rafters that will soon be headed off and trimmed back (top photo). In the bottom photo you can see how the two rafters at the top, now cantilevered over a new ridge beam, have been cut back at a taper to keep them out of the arched ceiling plane. To their right, an angled doubler picks up the weight of the ribs bearing on the base sheet. Benderboard strips backed by 2x blocking define the curve of the arched ceiling.

in. plywood. The short reverse-curve valley returns at each end were made of scabbed-on pieces of plywood (photo, p. 49).

The ribs are on 16-in. centers, and their bottoms are cut at an 8-in-12 pitch to match the roof slope (drawing, p. 49). As the individual ribs step up the roof, their overall depth decreases along with their width. Meanwhile, their arc at the top remains the same as that of the face frame. By taking direct measurements off the full-scale drawing, I got the overall depth of each rib. Then I measured down on the centerline of the window face frame to find the perpendicular baseline to read the width of each rib. To add a little extra complexity to the project, I had to increase the depth of the web of each rib in a sequential manner. This allowed the arched portion of the ceiling to make its transition into the cathedral ceiling without crowding the ridge (photo below).

I made all the parts for the rib-type dormer out of ¾-in. plywood. The face frame is a single layer, the built-up window frame has 5 layers and each rib has 3 layers. Rib number 1 is screwed to the back of the window frame.

Working with a full-scale drawing made for accurate and speedy work. But the pieces were large and cumbersome, and temporarily tacking them to the roof to figure out the shape of the base didn't sound like any fun at all. I probably could have used the full-scale elevation drawing to extrapolate its shape, but whenever I have to deal with unusual concepts, like sections of cones on inclined planes, I take comfort in three-dimensional models.

While regarding the cavernous interior of the recreation room, a solution occurred to me. Why not build a mockup of the roof, with one end of the rafters firmly planted on the recreation-room floor? Within an hour, I had a fake roof in place. I used the plywood that would eventually become the base sheet for its sheathing.

As each rib was cut out according to direct measurement takeoffs, I tacked its base to the fake roof and braced it plumb with a temporary alignment spine (photo, p. 49). Once I had all the ribs tacked to the mockup, I marked the inside and outside points where their bases engaged the plywood. These points gave me the reference marks I needed to make the bell-shaped base for the ribs. After taking the ribs down, I drove 8d nails at each mark, leaving enough of the nails exposed to act as stops. Then I used a ¼-in. by 1½-in. strip of straight-grained redwood benderboard held against the protruding nails to generate the curve for the base sheet.

The base sheet tucks into a bay between a pair of new timber-framed trusses (top photo, facing page). Unlike the installation of the rafter-type dormers, this one went into a roof that hadn't been planned with an eyebrow dormer in mind. This meant that some rafters had to be removed, and their loads picked up and transferred to new structural members.

Blocking and benderboard—Before I took apart the mocked-up structure of the dormer, I made a rosin-paper pattern to guide the cutting of the plywood sheathing. Like the pattern for the dormers on the garage, this one could be flopped to be the pattern for the other side of the dormer.

Assembling the ribs began from the bottom up. With the face frame and its accompanying arches firmly attached to the base sheet and diagonally braced plumb, all the succeeding ribs were quickly placed on their layout marks. They were then decked with a single layer of ½-in. plywood. As before, I built up skip sheathing over the top of the eyebrow with two layers of ⁵⁄₁₆-in. by 6-in. benderboard.

Picking up the loads of the removed rafters and carrying the curves of the ribs into the plane of the ceiling was the next task. Our engineer recommended using a couple of doubled 2x8s as support for the legs of the base sheet. These doublers run diagonally from the ridge beam to the top chords of the new timber-framed trusses (bottom photo, facing page). This photo also shows the finicky blocking that it took to pick up the unsupported edges of plywood sheathing and to carry the arc of the eyebrow into plane with the rafters. Benderboard was also useful for this task. In places, I was able to extend the curve from a rib to the rafters with a strip of benderboard and then fill in the remaining gaps with solid pieces of blocking shaped to fit.

The ends of the benderboard abut the edges of the drywall that cover the flat parts of the ceiling. At the transition to the curve, expanded metal lath was stapled over the benderboard, and the junction between the flat ceiling and the eyebrow's arch feathered with plaster to make an invisible seam. □

A plaster finish on the arched ceiling flows into the drywall covering the plane of the rafters.

James Docker is a building designer and general contractor living in San Carlos, California. Photos by author except where noted.

Framing a Hip Roof

After you've framed a gable roof, rafter templates and rafter tables
are all you'll need to make a hip

by Larry Haun

Fitting together pieces of the hip-roof puzzle. If all of the rafters have been cut properly, assembling a hip roof should be a painless process. Here, the author lines up a jack rafter for nailing.

I built my first hip roof in 1951 while in the Navy being trained as a carpenter. I dutifully laid out my rafters by stepping them off with a framing square. When I was finished, the commons were fine, but the hips came out short. Ever since then, I've relied on a book of rafter tables to determine rafter lengths rather than trust my ability to count steps with a square. Having framed hip roofs for so many years, I'm surprised that so many carpenters seem reluctant to build hip roofs. Maybe they're afraid that the framing is too complicated or beyond their abilities. I think that once you've learned to frame a gable (see the article on pp. 25-29), cutting and building a hip roof requires few additional skills.

A hip roof has the advantage of being inherently stronger than a gable roof. The hip rafters act as braces in the roof to resist the destructive forces of earthquakes, and the roof sloping up from all four sides of a hip roof offers no flat ends to catch high winds. Another advantage to hip roofs is that changing the roof style from gable to hip can transform the appearance of a house, offering a nice variation from the gable roof.

A hip roof begins with common rafters—
Hip rafters extend from the corners of the building up to the ridge. On both sides of the hips, common rafters, called king commons, meet the ridge at the same point as the hips (top drawing, facing page). The side and end king commons and the hip rafters are the main framing components of the hip roof.

The end king common runs from the middle of the end wall to the ridge. This rafter is the same pitch as the rest of the roof, and it is the key to the hip roof's ending with a pitched plane instead of the more common vertical gable. The hip rafters form the line of intersection between the side-roof and end-roof planes. The first step in framing a hip roof is determining the span of the roof, which establishes the location of the king commons. The garage featured in the photos in this article is 18 ft. 6 in. wide. The end king common, which is at the exact center of the span, is 9 ft. 3 in. from the outside of the garage. This number also represents the run of the rafters. After marking the location of the end king commons, I measure down the sides of the building the same distance, and then I mark the position of the side

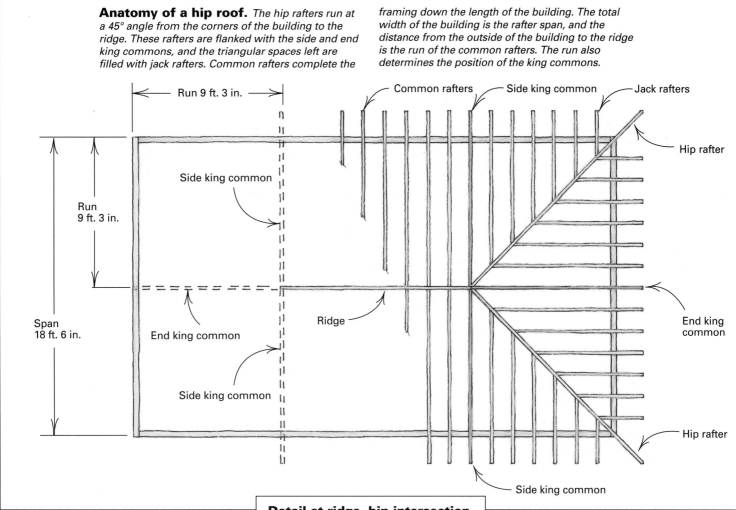

Anatomy of a hip roof. *The hip rafters run at a 45° angle from the corners of the building to the ridge. These rafters are flanked with the side and end king commons, and the triangular spaces left are filled with jack rafters. Common rafters complete the* framing down the length of the building. The total width of the building is the rafter span, and the distance from the outside of the building to the ridge is the run of the common rafters. The run also determines the position of the king commons.

- Run 9 ft. 3 in.
- Common rafters
- Side king common
- Jack rafters
- Hip rafter
- Run 9 ft. 3 in.
- Side king common
- Span 18 ft. 6 in.
- End king common
- Ridge
- End king common
- Side king common
- Hip rafter
- Side king common

king commons (drawing above). Next I lay out the rafter locations on the double-wall plates.

Rafter templates streamline measurement and layout—The roof of our garage has a 4-in-12 pitch, which means that the common rafters rise 4 in. vertically for every 12 in. they run horizontally. Because hip rafters run at a 45° angle to their neighboring commons in plan view, hip rafters must run 17 in. for every 4 in. of rise (drawing above). (By the way, 17 in. is the hypotenuse of a right triangle with 12-in. legs.)

When cutting rafters for any type of roof, especially a hip roof, rafter templates are a great way to speed the layout process. These neat little site-built aids have the rafter plumb cut on one end and the bird's mouth layout on the other. For this project I will need templates with pitches of 4-in-12 for the common rafters and 4-in-17 for the hips.

For the common-rafter template, I use a 2-ft. long piece of 1x6, which is the same width as my rafter stock (photos p. 55). I place my rafter square, or triangle square, on the template stock, pivot it to the correct pitch number (4) on the row of numbers marked "common," and mark the ridge plumb cut along the pivot side (photos p. 55). I then slide the square down the template about 1 ft. and make a second plumb mark for the heel cut of the bird's mouth. I square this line across the top edge of the template so that I

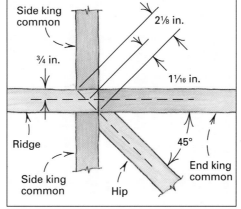

Detail at ridge, hip intersection
Because rafter length is measured from the center of the ridge, half of the thickness of the ridge must be subtracted. Our ridge is a 2x, so the commons have to be shortened by ¾ in. But the hips intersect the ridge at a 45° angle, so they must be shortened by 1¹⁄₁₆ in.

- Side king common
- ¾ in.
- 2⅛ in.
- 1¹⁄₁₆ in.
- Ridge
- Side king common
- Hip
- End king common
- 45°

can use the line as a reference when marking the rafters.

A level seat cut combines with the plumb heel cut to make up the bird's mouth of the rafter. The seat cut of the bird's mouth lands directly on the 2x4 top plate, so I make the seat cuts about 3½ in. long, squared off to the heel-cut line. The plumb distance from the seat cut of the bird's mouth to the top edge of the rafter is the height above plate and must be the same for both hip-rafter and common-rafter templates in order to maintain

the plane of the roof sheathing. Hip rafters are cut out of stock that is 2 in. wider than the commons so that the jack rafters will have full bearing on the hip. The hip template is also cut out of wider stock, in this case 1x8 (photos p. 55). The ridge cut is laid out the same as for the common-rafter template except that the square is pivoted to 4 and 17 if you're using a framing square or 4 on the hip-valley index of a triangle square. Again, I move the square down the template about 1 ft. and scribe a second plumb mark for the heel cut, with the line squared across the top. Next, I mark off the height above plate on the heel plumb line of the hip template and scribe the level seat-cut line at a right angle from this point.

Hip rafters need to be lowered at the seat cut—The height above plate for the hip rafters is measured from the centerline of the rafter. Because the two roof planes intersect at an angle, the top edge of the hip rafter needs to be beveled slightly from the centerline to maintain the roof planes (top drawing, p. 56). This process of beveling a hip rafter (or a valley rafter) is known as backing.

A more efficient solution to this problem is lowering the hip rafter slightly (called "dropping the hip") by simply cutting the seat deeper. The size of the drop depends on the thickness of the rafter

Bird's mouths are overcut. A wedge shape is cut out of each rafter to give it a place to land on the plate. These bird's mouths can be overcut just enough to remove the wedge.

All four hip rafters are laid out and cut at the same time. Short site-built sawhorses hold the rafter stock for layout and cutting. With all of the boards stacked together, only one set of measurements needs to be taken. Templates (photos facing page) do the rest.

Jack rafters are laid out four pairs at a time. Jacks oppose each other in pairs along both sides of the hip rafter. Each successive pair is shorter by a specified length than the pair above it. Diagonal marks remind the author to make his 45° cuts in opposite directions.

stock and the pitch of the roof. I determine this distance by using a framing square (drawing bottom right, p. 56). For this 4-in-12 pitch roof, I need to drop the hip about ¼ in.

I subtract that ¼ in. from the height above plate on my hip-rafter template and make a new level seat-cut line at this point. When my layouts are complete, I cut out the templates carefully to ensure their accuracy. After cutting the bird's mouth in the hip-rafter template, I rip the tail section to the same width as the common rafters, which allows the soffit material to be properly aligned. I finish the templates by nailing a 1x2 fence to the upper edge of the template.

The quickest way to get rafter lengths is from tables—All of the information needed to calculate rafter lengths is right there on any framing square. But out here in southern California, I don't know of any framers who still use one for this purpose. Some framers determine rafter length using a feet-inch calculator like the Construction Master (Calculated Industries Inc., 4840 Hytech Drive, Carson City, Nev. 89706; 800-854-8075). However, I prefer to get my figures from a book of rafter tables, such as *Full Length Rafter Framer* by A. F. Riechers (Box 405, Palo Alto, Calif. 94302).

First, I find the page in the book that lists the rafter lengths for a 4-in-12 pitch roof. The length of a common rafter for a span of 18 ft. 6 in. is listed as 9 ft. 9 in. The length of the hip rafter for the same span is 13 ft. 5¼ in. These distances are from the plumb cut at the center of the ridge to the plumb heel cut of the bird's mouth at the outside of the wall. If the calculation method is based on run instead of span, don't forget to split the span figure in half.

Because these lengths are figured to the center of the ridge, the actual rafter length has to be shortened by half the thickness of the ridge (bottom drawing, p. 53). For a 2x ridge, common

rafters must be shortened ¾ in., and because hip rafters meet the ridge at a 45° angle, they have to be shortened 1¹⁄₁₆ in. These amounts are subtracted from the rafter by measuring out at 90° to the ridge plumb cut.

Lay out rafters in stacks of similar lengths—Once all of the rafter lengths have been determined, it's time to lay out my stock for cutting. I usually use the house plans to get a count of the rafters, keeping in mind that a hip roof has an extra common rafter on each end. Using a pair of low site-built horses, I rack up all of the commons on edge with the crowns up. Next, I flush the ridge ends by holding the face of a 2x4 against the end of the rafters and pulling the rafters up to

it one at a time with my hammer claw. From the flushed end, I measure down my length on the two outside rafters, shortening my rafter measurement for the ridge. I snap a chalkline across the tops of the rafters as a registration mark for aligning the bird's mouth on the rafter template.

I then place my common-rafter template against the first rafter flush with the ridge end and scribe the ridge plumb-cut line. I slide this rafter to one side and continue down the line, leaving all of the rafters on edge. Next, I align the bird's mouth registration mark on the template with the chalkline on the rafters and mark my bird's mouth cutlines on each rafter. When all of the rafters are marked, I cut the ridges, moving the rafters over one at a time. Next, I flip the

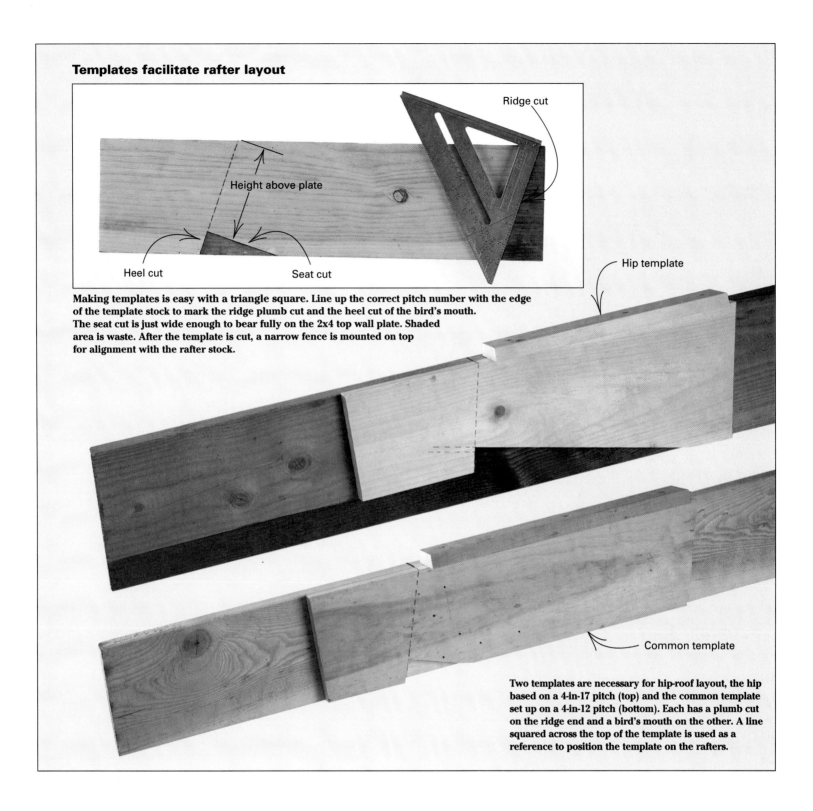

Templates facilitate rafter layout

Height above plate

Ridge cut

Heel cut **Seat cut**

Making templates is easy with a triangle square. Line up the correct pitch number with the edge of the template stock to mark the ridge plumb cut and the heel cut of the bird's mouth. The seat cut is just wide enough to bear fully on the 2x4 top wall plate. Shaded area is waste. After the template is cut, a narrow fence is mounted on top for alignment with the rafter stock.

Hip template

Common template

Two templates are necessary for hip-roof layout, the hip based on a 4-in-17 pitch (top) and the common template set up on a 4-in-12 pitch (bottom). Each has a plumb cut on the ridge end and a bird's mouth on the other. A line squared across the top of the template is used as a reference to position the template on the rafters.

rafters onto their sides and cut the bird's mouths, overcutting my lines just enough to remove the wedge without weakening the tail section (photo top left, facing page). I leave the rafter tails long and cut them to length after all of the rafters are in place.

A double cut brings the end of the hip rafter to a point—I try to pick out long, straight stock for the hips. The hip rafters need to be long enough to include the overhanging tail, which is longer than the tails on the commons. Most carpenters like to give hip rafters a double-side (or bevel) cut at the ridge so that they will fit nicely into the corner formed by the end and side king commons (bottom drawing, p. 53). I

do this by laying the hip-rafter template on the rafter stock and marking the ridge cut (drawing bottom left, p. 56). I then slide the pattern down 1½ in. and make a second mark parallel to the first. With my saw set at 45°, I cut along the first line in one direction and the second in the opposite direction, which leaves me a pointed end that will fit in between the king common rafters.

I set the hip stock on edge and flush up the pointed ridge ends (photo top right, facing page). Then I measure down from these points and make my plumb-heel-cut reference marks, shortening the rafters 1¹/₁₆ in. for the 2x ridge. Now the registration mark on my hip template can be aligned with the marks on my rafters, and I can scribe the bird's mouths.

To scribe the hip-rafter tails to the proper width, I hold a pencil against the tail part of the hip template and slide the template along the length of the tail. The bird's mouth of the hip rafters is cut just like the common rafters, and the tails are ripped to complete the cutting.

Jack rafters are cut in pairs—Jack rafters run parallel to the king commons and frame in the triangular roof sections between the king commons and the hip rafters. They are nailed in pairs into both sides of the hip rafter with each pair cut successively shorter as they come down the hip. The difference in length between each pair of jack rafters is constant (it's called the common difference), and it can be found in the rafter ta-

Two ways of dealing with hip rafters

The problem. *Without modification, the top edges of the hip rafter would be higher than the king commons.*

Dropping the hip. *The entire rafter can be lowered by deepening the seat cut. See drawing bottom right.*

Backing the hip. *The top edge of the hip rafter can be beveled slightly from the centerline to the outer edge.*

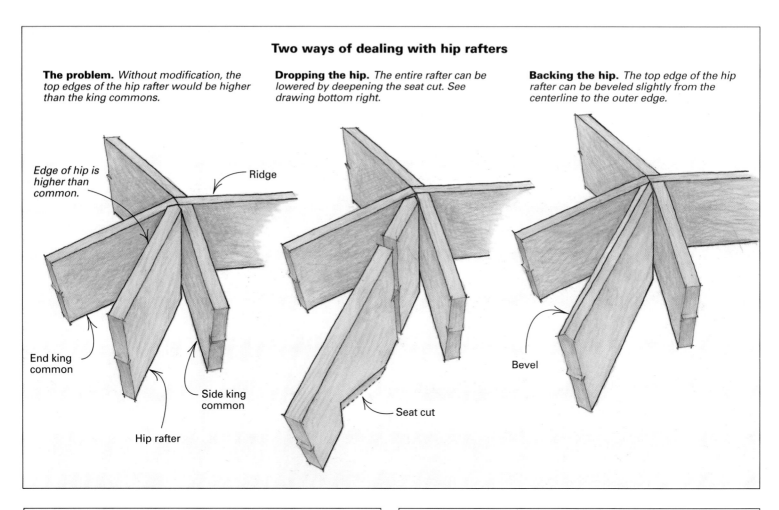

Edge of hip is higher than common.

Ridge

End king common

Side king common

Hip rafter

Seat cut

Bevel

The hip-rafter ridge cut. *A double side cut on the ridge end of the hip rafter lets it fit nicely between the side and end king commons. To make this cut, scribe two parallel plumb-cut lines from the ridge end of the hip template 1½ in. apart. With the sawblade set at 45°, saw along both lines in opposite directions.*

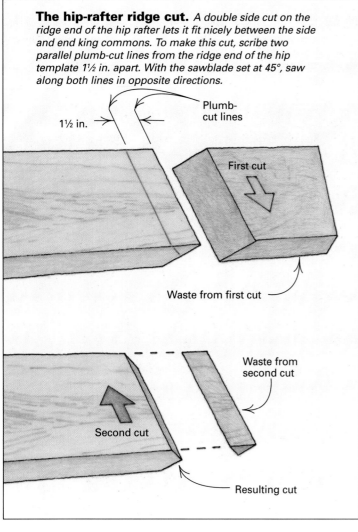

1½ in.

Plumb-cut lines

First cut

Waste from first cut

Waste from second cut

Second cut

Resulting cut

Finding the hip-rafter drop. *Hip rafters can be lowered slightly to put their edges in the same plane as the common rafters (see drawing below).*

Step 1. *Using a framing square, lay out a 4-in-17 pitch along the edge of any piece of rafter stock.*

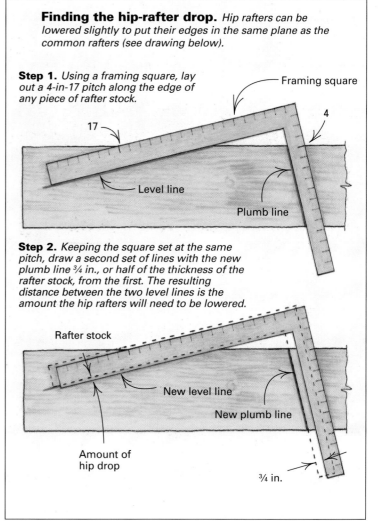

Framing square

17

4

Level line

Plumb line

Step 2. *Keeping the square set at the same pitch, draw a second set of lines with the new plumb line ¾ in., or half of the thickness of the rafter stock, from the first. The resulting distance between the two level lines is the amount the hip rafters will need to be lowered.*

Rafter stock

New level line

New plumb line

Amount of hip drop

¾ in.

bles. For jack rafters spaced at 16 in. o. c. at a 4-in-12 pitch, the difference in length is 1 ft. 4⅞ in. For 24-in. spacing, the difference is 2 ft. 1¼ in.

I lay out the jacks by racking together eight pieces of rafter stock the same width but slightly shorter than the common rafter (bottom photo, p. 54). (I rack eight pieces because there is a pair of jacks of equal length for each of the four hips.) Next to these I rack eight more pieces a foot or so shorter than the first eight and so on for each set of jack rafters. When the jack-rafter stock is laid out, I flush up the tail ends this time. The tails of the jack rafters are the same length as the tails of the commons, so I snap a line at that distance across all of the edges for my plumb heel cuts.

Next I lay an unshortened common rafter alongside my rack, lining up its heel-cut line with the heel-cut line on the jack stock. From the ridge cut of this common, I measure down the common difference. I shorten this first set of eight jacks by 1¹⁄₁₆ in. just like the hip rafter and make diagonal marks in opposite directions on each pair of jacks to remind me which way my cuts will go (bottom photo, p. 54). For each successive set of jacks, I measure down the common difference in length from the previous set. These measurements do not need to be shortened by the width of the hip rafter because I subtracted 1¹⁄₁₆ in. from my first measurement.

Using the common-rafter template, I mark the plumb side cut and the bird's mouth cut. Because pairs of jacks land on opposite sides of the hip, the 45° plumb cuts have to be laid out on opposite sides of each pair of rafters.

Assemble common rafters first—If everything is cut accurately, the roof members should fit together like a puzzle (photo p. 52). I always tack down plywood sheets on top of the ceiling joists for a safe place to work. The ridge length and rafter layout can be taken directly from layout on the wall plate, but I prefer to bring up a ridge section and begin my rafter layout about 6 in. from one end. I like having this extra length to compensate for any discrepancies in my layout.

With a partner, I set up my first pair of side king commons and nail them to the plate and into the ceiling joist. (In high-wind areas, rafters may need to be tied to the plates with a metal framing anchors.) Next, I go to the other end of the ridge and nail in another set of commons.

The ridge board then gets pulled up between the two sets of commons and nailed in place. I just tack the side king commons to the ridge until the end common has been installed. At this point I make sure the ridge is level by measuring from the tops of the ceiling joists at each end. I support the end of the ridge with a 2x4 leg down to a ceiling joist or to an interior wall and run a diagonal sway brace to keep everything in place temporarily. Next, I slide one of my side king commons out of the way, hold the end king common next to the ridge and mark the end of the ridge. After the ridge is cut to length, I nail my end king common in place.

Next, a hip rafter is toenailed to the wall plate directly over the outside corner. The side cut on the ridge end gets nailed to the end common next to the ridge. I nail the opposing hip in place,

The hips and king commons come together at the ridge. The end of the ridge is the meeting point for all of the major framing members of the hip roof. After the end king common is nailed in, the hip rafters are installed, and the ends of the side king commons are nailed in next to the hips.

and the two side king commons can be slid back against the hips and nailed in permanently.

If the roof is long, additional ridge sections may be installed using other pairs of common rafters for support. Again, I make sure additional ridge sections are level. At the other end of the building, I mark and cut the ridge and assemble the hips and side king commons as at the first end.

Frieze blocks stabilize the rafters—Before nailing in the jack rafters, I sight down the hip rafter and make sure it is straight from the ridge to the plate. If it's bowed, I brace it straight temporarily until the jacks are in. I start with the longest pair of jacks and nail them to the plate along with frieze blocks, which are nailed in between the rafters at the plate (photo p. 52).

Local codes don't always call for frieze blocks, but I use them to stabilize rafters and provide perimeter nailing for roof sheathing. If necessary, they can be drilled and screened for ventilation and are a good way to use scrap lumber. I cut a bunch of blocks ahead of time to either 14½ in. or 22½ in., depending on my rafter spacing.

Frieze blocks can be installed flush with the wall, where they serve as backing for the exterior siding. However, with this method the blocks need to be ripped to fit below the roofline. Another method is installing the blocks perpen-

dicular to the rafter just outside the plate line. I like this second method because it requires no ripping and provides a stop for the top of siding.

I nail in the frieze blocks as I install the remaining pairs of jack rafters. Each jack is nailed securely to the hip rafter; I take care not to create a bow. Once all of the pairs of jacks are installed, the hip will be permanently held in place.

The corner frieze blocks get an angled side cut to fit tight against the hips. Once all of the jacks and commons are nailed in, the rafter tails can be measured, marked and trimmed to length. Remember to measure the overhang out from the wall and not down along the rafter. For this building the overhang is 20 in., and the fascia stock is 2x (1½ in. thick). I mark a point on the top edge of the rafters 18½ in. straight out from the walls at both ends of the building and snap a line across the rafters between my marks. I extend my chalklines out over the tails of the hip rafters to mark the overhang at the corners. When marking the plumb cut on the rafter tails, use the common-rafter template on the commons and the hip template on the hips. ☐

Larry Haun of Los Angeles, California, is the author of The Very Efficient Carpenter, *a book and video series published by The Taunton Press. Photos by Larry Hammerness except where noted.*

Framing a Bay Window With Irregular Hips

How one carpenter calculates the tough cuts

by Don Dunkley

My crew and I frame houses in central California, near Sacramento, where designers compete with one another to see who can create the most complicated roofs. To stay in business, the local carpenters have to be adept at framing every type of roof—hip, gable, octagon, cone—sometimes all in the same building.

One modest and enduring feature that turns up in many of these homes is the bay window popout. The kind we build most frequently, and the subject of this article, consists of two 45° corners and a projection, or offset, of 2 ft. (floor plan, p. 60). It is 10 ft. wide at the wall line and 6 ft. wide at the front of the offset. The plate height of the bay and of the adjoining room are 8 ft. 1 in., and the roof pitch is 8-in-12.

A hip roof commonly tops this kind of a bay. But unlike many of the hip-roof bays that get built locally, we frame ours with two irregular hips (photo above right). More often than not I run across roof plans that leave out a second irregular hip. Without it, the plane of the roof has to be warped to intersect the valley (photo above left). Once you become aware of this refinement, chances are you'll spot many an example of incorrectly framed bays on a casual drive down a residential street. Adding the second irregular hip allows the roof planes to meet at crisp angles.

Building a roof with one pair of irregular hips is a challenge—add another pair and it's a task for a journeyman carpenter. When I first

started out as a framing carpenter, I spent a lot of time laying out the rafter locations on the subfloor, then transferred them by way of plumb bobs and stringlines to temporary staging, where another session with stringlines and tape measures would follow as I puzzled out seat cuts and cheek cuts. No more. I've incorporated two tools into my roof-cutting procedures that do away with all the plodding.

The first is a Construction Master Dimensional Calculator (Calculated Industries, Inc., 22720 Savi Ranch Pkwy., Yorba Linda, Calif. 92686). This calculator works in decimal numbers and in feet and inches. It also has pitch, rise, run, diagonal and hip/valley functions, which eliminate some of the key strokes required to apply standard calculators to carpentry work. For instance, instead of using a square-root formula to find a diagonal, I enter 11 ft. as rise, 14 ft. as run, punch the diagonal button, and the calculator will read 17.8044. When I punch the convert-to-feet-and-inches button, it tells me 17 ft. 9⅝ in.

The second tool is a book called *Roof Framing,* by Marshall Gross (Craftsman Book

Leaving the second irregular hip out of a bay-window roof causes an awkward warp in the sheathing, which shows up in the shingling and the valley flashing (photo above left). The bay in the photo at the right was framed with both hips, and the valley runs straight and true.

Company, P. O. Box 6500, Carlsbad, Calif., 92008). Gross uses a technique to lay out roofs that he calls the "height above plate" method (HAP). Simply stated, the HAP system allows me to set the ridges first at their actual height. Then I bring the rafters to meet them. I've found this system to be unbeatable for assembling complex roofs. But before we dive into HAP and bay window/roof theory and calcs, let's look at layout and walls.

Patterns and plates—We build bays like the one shown in the photograph (top photo, facing page) on either slab or wood-framed floors, and sometimes on cantilevered joists. In each case, laying out the bay begins after the subfloor is in place and I've snapped a chalkline marking the inside edge of the wall plates around the house.

To save time and ensure accuracy when I mark the position of the bay, I use a plywood pattern of a 2-ft. by 2-ft., 45° corner (floor plan, p. 60). By placing the pattern along the wall line at the beginning of the bay, I can quickly lay out a perfect 45° wall. The pattern has layout marks on both sides, so I can flip it to lay out the opposite corner.

I usually make square cuts on the ends of the diagonal wall plates. They abut outer wall plates that have two 45° cuts on their ends (floor plan detail, p. 60). I do this for two reasons: the angled stud on the outer wall plate gives me a good nailing surface to anchor the walls together, and it gives me a little more room to squeeze the window into the diagonal wall. Designers inevitably want lots of windows in these walls. That means I need from 26½ in. to 27½ in. for my header, depending on the width of the windows. As you can see from the photo below, this can get snug. To make the windows fit, I sometimes have to use 1x4 king studs instead of 2x4s next to the trimmers that carry the window headers.

After the walls have been framed and plumbed, it's time for the roof. If you're familiar with roof theory, I'll go straight to calculating the rafters for the bay. If you'd like to brush up on roof basics, please refer to the sidebar, "Regular and irregular hips," on p. 63.

Locating the ridge—To find the ridge height using the HAP method, I add the distance the rafter sits above the plate at the seat cut to the theoretical rise, minus the reduction caused by the thickness of the ridge (elevation view, p. 61). For example, our seat cut (the horizontal portion of a bird's mouth) is 3½ in. on the level, giving a 4¼-in. rise above the plate for a 2x6 rafter. The run of the common rafter is 5 ft., and the rise is 40 in. (8-in. pitch by 5-ft. run). This gives us a theoretical rise of 44¼ in. If there were no ridge, the peak of the rafters would be this height, but the ridge comes between them and must be accounted for. This applies to both common rafters intersecting the ridge at a right angle or, as in this case, a common rafter in line with the ridge. I find the reduction the "new-fashioned" way, courtesy of Construction Master (detail 1, p. 61). Our ridge is 1½ in. thick. Using my Dimensional Calculator, I enter half the thickness of the ridge (¾ in.) as the run. Next I enter 8 in. as the pitch, and punch the rise button. My answer is ½ in. I subtract that from 44¼ in. to get the actual height of the ridge above the plate: 43¾ in.

To begin erecting the bay's roof, I set its ridge on temporary legs so that it sits precisely 43¾ in. above the plate. If I later cut all my rafters accurately, and my walls have been properly plumbed, lined and braced, all the parts will converge to lock the assembly together.

Because the offset of the bay is 2 ft., the ridge is approximately 2 ft. long. Actually, it's a little longer in order to compensate for the shortening allowance of the common rafter. More on this in a minute.

First, the valleys—Our floor plans show a 2-ft. offset and a 6-ft. long front to the bay window. The interior opening of the bay is 10 ft. wide. The roof plan shows two valley rafters, one common rafter and two sets of irregular hip rafters (roof plan, next page). The roof overhang is 2 ft.

At this stage of the roof framing I'm not concerned about the 45° wall of the bay. Instead, I'm thinking about the 10-ft. wide

Drawings: Michael Mandarano

opening to the bay. Dividing it in half gives me a pair of 5-ft. squares in plan. The bay's ridge and common rafter form a line between the squares, and its regular valleys are the diagonals.

An 8-in-17 valley (see sidebar) on a 5-ft. run calculates to be 7 ft. 9¾ in. long. To get my 2-ft. overhang, I have to add 3 ft. 1½ in. from the seat cut to the tail cut. The vertical edge (heel cut or plumb cut) of the bird's mouth aligns with the exterior face of the wall framing.

The skinny little lines we see when we look at roof framing plans represent the center lines of rafters and beams. To transfer the ideal of a line with no width into a rafter that is typically 1½ in. thick, we have to take the shortening allowance (SA) into consideration. For a regular hip or valley, the SA is equal to half the thickness of the common rafter, cut on a 45° angle (detail 2, p. 87). That works out to 1¹⁄₁₆ in. for 2x framing lumber. Remember this is a level measurement, and has to be adjusted for the pitch of the roof. For our 8-in-17 pitch, the valley rafter has to be shortened by 1³⁄₁₆ in. To find the adjusted SA with the Construction Master, enter the pitch as the decimal .47 (8-in. rise divided by 17-in. run). Then enter 1¹⁄₁₆ in. as the run, and punch the diagonal button to get the adjusted SA of 1¹¹⁄₆₄ in. Round it off to 1³⁄₁₆ in.

I'm in the habit of cutting double cheek cuts on valley rafters. Often there are other rafters intersecting the same ridge, and the double cheek cut gives me a little extra room for adjustment. In addition, to save

time at the cutting table I put double cheek plumb cuts on all my valley and hip stock at the same time, and then decide later which ones end up as hips or valleys. The photo on p. 62 shows how the valleys intersect the ridge.

Common rafter—In order for the common rafter to be at the same height as the valley rafter, it also must be calculated on the 10-ft. span, or 5-ft. run. An 8-in-12 common on a 5-ft. run calculates to be 6 ft. ⅛ in. long. Measured on the level, its SA is half the

The author nails down irregular hip rafter A on a cantilevered bay (photo below). Note the 1x4 king studs in the diagonal walls. They allow more room for the window. Opposing walls are joined by a tie beam across the top of the opening to the bay (photo above). The upper half of the top plate has been let into the beam for several feet.

thickness of the ridge—in this case ¾ in. (detail 1, facing page).

Our next move is to calculate the lengths of the valley jacks. Since our run from the common rafter to the valley rafter is 2 ft., our valley jack will be cut on that run. The valley jack will have two SAs—one for half the thickness of the ridge and one for half the thickness of the valley measured across its top at a 45° angle.

Irregular hip A—Looking at the roof plan (bottom drawing) we see that the distance from the common rafter to irregular hip A is 3

ft. The run of the common rafter to the ridge is 5 ft. To find the run of hip A, I enter 3-ft. rise and 5-ft. run. When I punch the diagonal button, it reads 5 ft. 10 in. But that's measured on the level. To figure the length of hip A, we need its actual rise. By feeding our 40-in. rise and 5-ft. 10-in. run into the calculator, I get the full unadjusted length of the rafter from seat cut to ridge junction: 6 ft. 8⅝ in.

While we're working with these numbers, let's figure out the plumb cut for this hip by dividing our 40-in. rise by our 5-ft. 10-in. run to get tangent .571428. The trig tables say

that's almost 30°. While the tangent number is still in the calculator's display screen, I punch the convert-to-inches button, which now reads 6⅞ in. That means the pitch (and the plumb cut) of this hip rafter is 6⅞-in-12.

Because we're dealing with an irregular hip here, we need to know the angles formed by its intersection with the plate and the common rafter. Without them we can't calculate the cheek cuts on the hip rafter at the ridge or the cheek cuts on the hip jack rafters. Using the tangent method and the trig tables, I find that the angle made by the hip

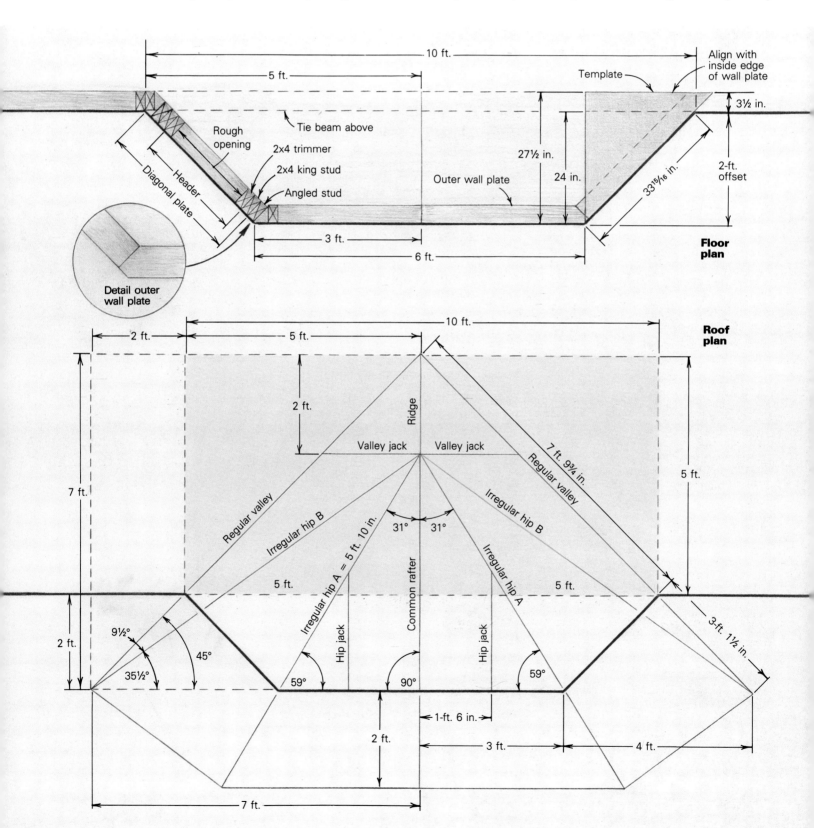

and the plate is 59°; therefore the complementary angle is 31°.

To be an irregular hip is to be off center at the intersection of all the other rafters (plan of rafter intersection, next page). Here's a way to calculate the SA and cheek-cut angles for this asymmetrical junction. In the triangle ABC, the rise of BC is half the thickness of the ridge, or ¾ in. The angle A is 31°, derived from the plan view of our roof. Thirty-one degrees is the same as a 7³⁄₁₆-in. roof pitch. Calculator in hand, I enter 7³⁄₁₆ in. as pitch and ¾ in. as rise, punch the diagonal button, and it reads 1⁷⁄₁₆ in.

for the hypotenuse AB in triangle ABC. This is the SA, measured on the level. The adjusted SA for this 6⅞- and 12-pitch is 1⅝ in.

Let's take a look at the rafter intersection plan to see how the framing square is used to lay out this irregular-hip plumb cut. First, measure back from the full length of the rafter the adjusted SA, 1⅝ in., to mark point X on the rafter's centerline. By looking at the plan view we see that the hip needs two different cheek cuts. Let's make the longest side first. By laying the framing square on top of our rafter with the tongue set at 7³⁄₁₆ and the body set at 12, draw

a line on the edge of the rafter that passes through point X (detail 3, next page). This line (YD) gives us the angle for the cheek cut on the side of the common rafter.

By studying our plan of our rafter intersection, we see that the two cheek cuts intersect off the centerline of the rafter at point F. To find this point, first square a line from the edge of the rafter to X to find point Z. A line perpendicular to YD that intersects point Z gives us the second cheek cut in plan.

Now we've got the cheek-cut angles for a horizontal rafter. Just to make this a chal-

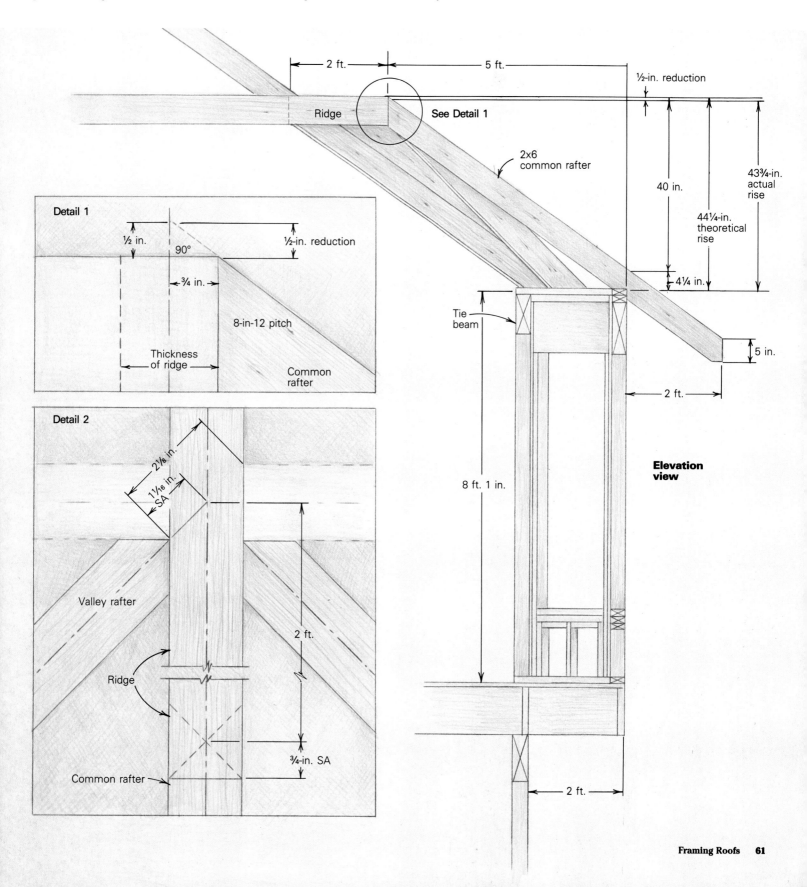

Converging rafters meet at the end of the bay's ridge beam. In the lower left corner of the photo you can see where the valleys intersect the ridge. The "X" marks one end of the 2-ft. ridge.

lenging exercise, the angles change as the rafter's pitch increases—the greater the pitch, the greater the change. You can demonstrate this phenomenon by drawing an equilateral triangle on a slip of paper. Hold the drawing level, with its base toward you. Now slowly rotate the drawing to vertical to change its pitch. You can watch the angle change from an obvious 60° to a right angle and beyond.

When you make compound cuts with a circular saw, such as the cheek cuts on a hip or jack rafter, the saw automatically compensates for the pitch of the rafter. But if you have to use a handsaw to cut an angle beyond the circular saw's 45° capability, you have to compensate for the angle change in your layout. The angle we need to cut here is 59°. Here's how to lay it out.

Recall that our hip plumb cut is 6⅞ and 12. Scribe a line at this pitch on the side of the rafter, beginning at point D (detail 4 below). Now mark from this line the distance EY on the side of the rafter, and scribe another line at 6⅞ pitch. Square this line across the top of the rafter to find point Y1. Connect point F and point Y1 to find the adjusted cheek cut. If you want to cut the other angle with a handsaw, repeat the process to find the adjusted cut for the other cheek. In plan, they look like detail 5 below. Now you're ready to make the double-cheeked plumb cut for irregular rafter A, and to take a break.

In practice, when the rough framing is going to be covered up by a ceiling, I generally trim this cut to fit. A perfect cut isn't necessary for structural integrity, and as always, time is of the essence. But a journeyman carpenter should know how to make this cut precisely if the rafters are going to be exposed to view.

I let the tails of these two rafters run wild past the wall. Once I've got the rest of the rafters in place, I use my level to determine the position of my tail cut. In this manner I can make sure that the fascia and gutters end up in the right place.

Irregular hip B—These hips can be calculated mathematically, but to tell the truth I

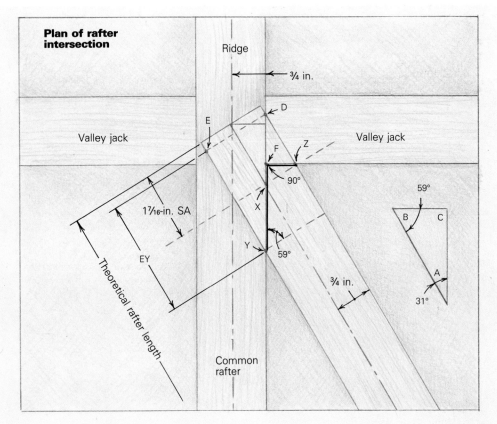

Plan of rafter intersection

Ridge

¾ in.

D

E

Valley jack

Valley jack

F

Z

90°

59°

B   C

1⁷⁄₁₆-in. SA

X

EY

Y

59°

¾ in.

A

31°

Theoretical rafter length

Common rafter

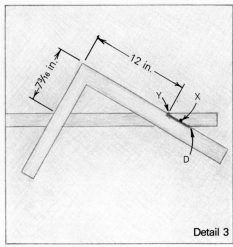

7³⁄₁₆ in.

12 in.

Y   X

D

Detail 3

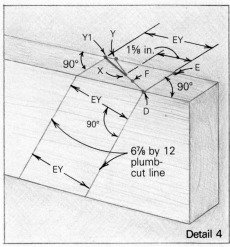

Y1   Y

EY

90°

1⅝ in.

X   F

E

90°

EY

90°

D

EY

6⅞ by 12 plumb-cut line

Detail 4

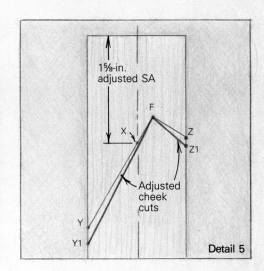

1⅝-in. adjusted SA

F

X

Z

Z1

Adjusted cheek cuts

Y

Y1

Detail 5

use stringlines to figure them out. I tack a nail in the top of the valley rafter to represent the centerline intersection of the valley and its neighboring hip. Then I run a string to the point at which the rafters are intersecting at the end of the ridge. I'll measure this distance to get the unadjusted length of the rafter, and while I'm at it I measure the distance from the stringline to the wall plate to get the depth of the seat cut.

To lay out the radically tapered tail cut on hip B, I need to know the angle the rafter makes with the front of the bay. Using the tangent method, I find it to be 35½°. From the plan view we see that the 45° valley and 35½° hip come together to form a 9½° angle (45° minus 35½°). The complement of 9½° is 80½°. I use this angle on my Speed Square to mark the tail cut. Once I adjust the cheek cut for the pitch of the rafter, I make the cut with a sharp handsaw. The only rafters left to install are the hip jacks.

Because I have 3 ft. from corner to common rafter, I center the jack at 1 ft. 6 in. You should have enough information now to figure out their rise, run, pitch, length and cheek cuts. □

———————
Don Dunkley is a framing contractor working in California's central valley.

Regular and irregular hips

Understanding complex roofs requires an understanding of the mathematics of a simple roof. Here are some basics.

The *pitch* of a roof is determined by the relationship of vertical *rise* to the horizontal *run*. An 8-in-12 roof means that for every 12 in. of run the rafter will rise 8 in. To represent this relationship visually, a diagonal line connects the two, forming a triangle (detail C below). The diagonal represents the slope of the rafter. Because the rise and run are perpendicular to one another, the three lines form a right triangle. The mathematical formula to find the length of the diagonal of a right triangle is called the Pythagorean theorem: $a^2 + b^2 = c^2$. Another way to figure it: c = the square root of $a^2 + b^2$

Fortunately, the carpenter can reach for a calculator to process these numbers. Another way to bypass tedious rafter calculations is with a rafter book, such as *Full Length Roof Framer* (A. F. Reichers, Box 405, Palo Alto, Calif. 94302). It lists the lengths for common, regular hip and valley rafters for roofs with 48 different pitches.

If you divide the rise by the run, you get the *tangent*. For an 8-in-12 roof, the tangent is .666667. What can you do with this information? By looking at a table of trigonometric functions you'll find that a tangent of .666667 is equal to approximately 33¾°. Therefore an 8-in-12 roof rises at an angle of 33¾°. All roof pitches have a corresponding tangent/degree.

Once we know two of the angles in a triangle, we can subtract their sum from 180° to get the third angle. In our example, 180° - (90° + 33¾) = 56¼°, which is called the *complement* of the 33¾° angle (complementary angles add up to 90°). When laid out with a Speed Square, this 56¼° angle will give the *level* or *horizontal cut* of a rafter whereas the 33¾° is the *plumb cut* (detail A).

In plan, a *regular hip* rafter intersects common rafters at a 45° angle. To understand a hip, look at it from the plan, or top view (drawing below). The common rafters intersect the plates at 90°, while the hip is 45° to the plates. In order for the regular hip rafter to reach the same point as a common rafter, its run must be longer. For every 12 in. a common rafter needs for run, a regular hip rafter, regardless of pitch, needs 16.97 in. (for pitch designations, carpenters round the number off to 17 in.). This relationship holds true for *regular valley* rafters as well. Therefore, a regular hip or valley rafter on an 8-in-12 roof is cut to a 8-in-17 pitch. Divide 8 by 17 to get the tangent: .4705, which gives us the hip *plumb cut* — 25½°.

As shown in our plan, a regular-hip roof over a 22-ft. span reveals two 11-ft. squares. The run of the commons will be 11 ft. Find the run of the hips by multiplying the run of the common times the run of a regular hip: or, 11 x 16.97 in., which equals 15 ft. 6¹¹⁄₁₆ in.

So what does this tell us about how to calculate irregular hips? To figure out an *irregular hip* the diagonal length of its run is needed. But because an irregular hip doesn't have a 45° angle in plan, the value of 16.97 can't be used. Back we go to the Pythagorean theorem.

Let's say we have an 8-in-12 roof with commons that run 11 ft., but the commons are 14 ft. from the corner where the hip originates at the plates (drawing below and detail B). By using the Pythagorean theorem, we find that the diagonal in a triangle with 11 ft. and 14 ft. sides is 17 ft. 9⅝ in., which gives us the run of this irregular hip. To find the full length (also referred to as theoretical or unadjusted length) of the hip rafter, use the rise (88 in.) and the run to find the diagonal, which is 19 ft. 3¹⁄₁₆ in.

The hip jacks are another wrinkle — those intersecting an irregular-hip rafter have different angles on their cheek cuts. Using our tangent formula, we find that our irregular hips divide the plan view of our roof into 38° and 52° angles. When the cheeks of opposing jack rafters are cut at these angles and adjusted for the pitch of the rafter as shown in the bottom drawing, p. 60, they'll fit snug against the irregular hip. —*D. D.*

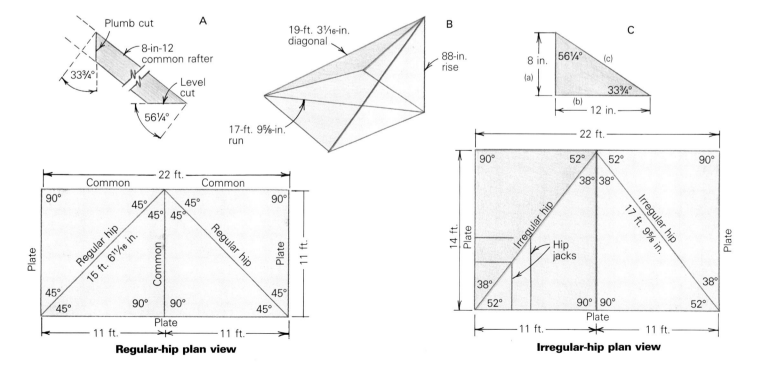

A

Plumb cut
8-in-12 common rafter
33¾°
Level cut
56¼°

B

19-ft. 3¹⁄₁₆-in. diagonal
88-in. rise
17-ft. 9⅝-in. run

C

8 in. (a)
56¼° (c)
33¾°
(b)
12 in.

Regular-hip plan view

22 ft.
Common   Common
90°   45° 45°   90°
     45° 45°
Plate   Plate
Regular hip   Regular hip
15 ft. 6¹¹⁄₁₆ in.
Common
45°   45°
90° 90°
Plate
11 ft.   11 ft.
11 ft.

Irregular-hip plan view

22 ft.
90°   52° 52°   90°
       38° 38°
14 ft.
Plate   Irregular hip   Irregular hip   Plate
                17 ft. 9⅝ in.
Hip jacks
38°   38°
52°   90° 90°   52°
Plate
11 ft.   11 ft.

Ceiling Joists for a Hip Roof

Three simple, problem-solving framing techniques

by Larry Haun

Hip rafter

Header

Blocking

Plate

Heading out the joists

Framing models by Linden Frederick

If you want a simple roof, go for a gable. The angles are basic, and even the ceiling framing is a snap because it generally runs parallel to the rafters. A hip, however, is a roof to reckon with, and not just because of its angles. With rafters sloping up from all four sides of the building, how do you joist the ceiling?

"No problem," you say. But what about the joists that run perpendicular to the rafters on opposing sides of the house? Depending on the roof's slope, it's quite likely that you won't have much room for the joist closest to the plate: the rafters will be in the way. In 41 years of pounding nails, I've seen several solutions to the problem of joisting a hip roof.

Time-honored solutions—Depending on the size of the joists, the ones closest to the plates may need to end at a header to let the hip rafter through (photo left). A similar header is sometimes required for the valley rafter. Backing for the ceiling finish is obtained by nailing 2x blocking flatwise to the plates between the rafters (behind the frieze blocking), to provide 1½ in. of nailing surface for the ceiling drywall.

There are times, however, when the first joist may have to be held away from the plate as much as 32 in., depending on the pitch of the roof and the size of the lumber used for the joists. The traditional method of filling the space between the outside wall and this first joist is to add stub joists at right angles to the main joists, parallel to the jack rafters (bottom photo, facing page). This technique often requires that the first joist be doubled to carry the extra ceiling load. The stub joists are usually clipped on one end to keep them from sticking up above the roof framing; pressure blocks at the other end help to support them.

A better solution—There's another way to handle the problem, one that requires less material and labor. Set your last joist as close as possible to the plate. Then set another one—flatwise—within 16 in. of the outside wall. There will be plenty of room between it and the underside of the hips and jacks. The next step is to install the rafters and to nail frieze and backing blocks between them as described earlier. Then cut strongbacks from scrap 2x stock and run them flatwise from the plate to the first on-edge joist (top photo, facing page). You won't need many of them; spacing them about 4 ft. o. c. should do.

Secure the strongbacks to the backing blocks and to the upright joist with a couple of 16d's at each end. At the joist end the strongbacks must be held up 1½ in. from the bottom. Finally, pull the flat joist up to the strongback, securing it with three or four more 16d's, angling them slightly for better holding power. The strongbacks will stiffen and support the flatwise joist. With this step complete, the ceiling is ready and you're set to move on. □

Larry Haun lives in Los Angeles and is a member of Local 409; he taught in the apprenticeship program. Photos by Susan Kahn.

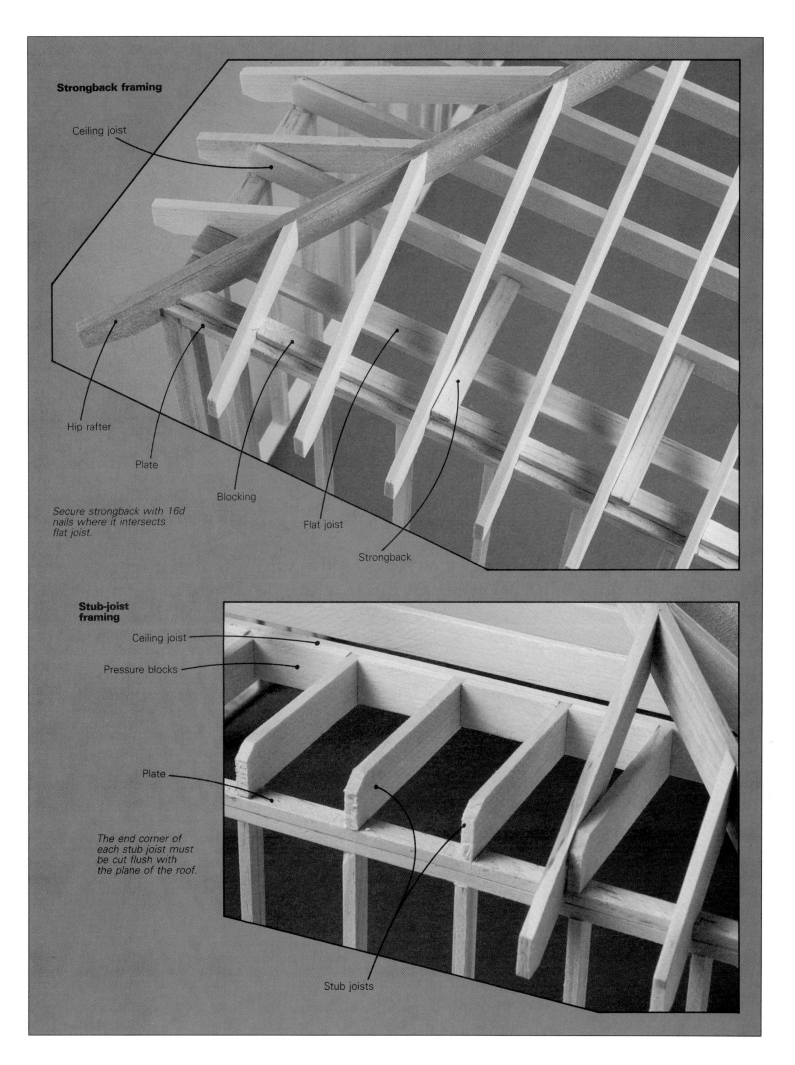

Strongback framing

Ceiling joist

Hip rafter

Plate

Blocking

Flat joist

Strongback

Secure strongback with 16d nails where it intersects flat joist.

Stub-joist framing

Ceiling joist

Pressure blocks

Plate

The end corner of each stub joist must be cut flush with the plane of the roof.

Stub joists

Pagoda Roof Framing

Tapered blocks at the corners of this hip roof lift the eye heavenward

Easier than it looks. The upswept corners of a pagoda roof are not difficult to build once the hip roof has been framed. Three curved pieces of 2x material added to the corners give the roof its distinctive shape. Aluminum roof panels flex enough to fit the contours.

by Tim Carr

The Oriental flavor and natural lines of the pagoda roof are uplifting to the spirit, like a man with arms extended, palms up, in a universal gesture of peace and welcome. In comparison, most two-story buildings have a boxy, top-heavy look, with little appeal in their proportions or lines. My decision to build a pagoda in Hawaii and experiment with its potential came from a desire to blend the building with the islands' flowing, casual lifestyle. The lines of this little studio and library are a mix of some of the best and most-pleasing ethnic and cultural expressions found throughout the islands.

The site for this pagoda (photo above) is a steep hillside overlooking a 30-ft. waterfall and pool. The setting gives the building its inspiration as a music room, library, architectural design studio and workshop. The gently upswept roof cor-

ners are important, but they weren't hard to add after the hip roof had been framed conventionally. This kind of roof is simple enough for the amateur builder and should be a pleasure for professionals who want to create a major design statement easily. Although the technique is straightforward, it helps to have a bandsaw and a compound-miter saw on hand to cut the curved pieces that form the corners of the roof.

Building up roof corners—What gives the roof corners their upswept shape are three extra pieces of 2x material that are applied to standard roof framing (drawings facing page). The curves of all three pieces rise together to meet at each corner, lifting the lines of the fascia and the hip. Two of the three pieces have identical shapes—they are really mirror images of each other—and

are applied on both sides of each corner right above the fascia. A third curved piece, the center support, sits on top of the hip rafter. I experimented with the shape of this roof after the basic roof framing was done, and the redwood fascia had been installed. By varying the shape and dimensions of the three extra pieces, I could alter the height and the balance of the corner profile. In the end I chose a center support with a rounded outer edge to complement the half-round gutters I had picked. On this building, which is 20 ft. by 24 ft., the 8-in. height and 40-in. length of the curved fascia pieces looked balanced. Each center piece is 44 in. long and 9½ in. high at the outside corner, where it extends slightly past the fascia pieces. To me, the overall effect is similar in feel to the Golden Temple in Kyoto, one of Japan's most famous buildings.

I nailed plywood gussets—8-in. by 4-in. pieces of ¾-in. plywood—on both sides of the center supports to attach them to the hip rafters. The gussets extend below the bottom edge of each support so that they straddle the hip rafters and form a kind of saddle. The gussets helped hold the center supports in place while I was trying to align them and provided additional nailing area to attach the supports to the roof. If you are not adding a soffit, I recommend deleting the gussets for a cleaner look.

Each center support requires two 2x2 nailers, applied to the upper edge on each side. The nailers are cut to the same curved profile as the support and have 15° bevels on one side. The beveled edges are nailed to the center support, so the nailers angle slightly downward. The angle ensures a flat nailing surface for the roof panels that sweep up to the support. The nailers terminate where they meet the curved fascia pieces and, if laid out properly, will make stops for nailing all three pieces together. The attachment of the fascia curve also was made easier by an additional 1x1 strip nailed to the bottom inside edge of each fascia piece. The strips could then be nailed to the 2x4 purlins that sit on top of the fascia boards.

Templates for roof panels—After the framing was complete, and the additional corner pieces were in place, I needed to find a way to shape the aluminum roof panels to fit the curves of the roof. Instead of trying to cut the aluminum panels in place, I made templates first, using ¼-in. plywood, then transferred the shape to the roofing panels. I screwed the plywood down to the roof framing so that it conformed to the curves, running the long edges of the sheets parallel to the jack rafters (the hip rafters ran beneath the panels diagonally). At the bottom edge, I left a 1½-in. overhang beyond the fascia so that water running off the roof would be directed into the gutters. The plywood had to be scribed along the edge of the center support before the curved nailers were attached. Because I was scribing along the far edge of each 2x center support, I adjusted the line back ¾ in. so that the roof panels on both sides of the support would meet in the middle. I used a black felt-tipped marker with a wide point, so I got a good nonreflective line to follow while cutting. After cutting the template with a sabersaw, I double-checked the fit before cutting out the aluminum. In cutting this template, you might expect a concave line to result, but it came out convex in a compound curve.

The roofing I picked, made by Reynolds Aluminum, is common in Hawaii. Its diamond-rib pattern adds to its structural stability; it's also light, easy to work, fireproof and recyclable. The material is easy to cut with a circular saw equipped with a metal-cutting blade, but I still wear a face shield, long sleeves and ear protection. I cut the roof panels at ground level, where I could set up sawhorses and supports and have plenty of room to work without having pieces fall off the roof. When installing the panels, I started at the right corner of each roof section and worked toward the left, which made nailing much easier and kept me off the aluminum as I worked my way around the building.

Blocks make the difference. *This pagoda roof begins with a conventionally framed hip roof. Extra 2x material—a curved center support and two curved fascia pieces—is added at the corners to give the roof its shape. Curved nailers are cut to 2x2 dimensions, beveled on one edge and attached to the curved center supports, giving a flat nailing area for the roofing panels.*

Plan view

Hip rafter
Nailers
2x4 purlin
Curved 2x2 nailers
Curved fascia
2x6 jack rafters
Curved center support
2x fascia below
Plywood gusset
Curved fascia

Section at hip rafter

2x4 purlins
2x4 nailer
Curved 2x2 nailer beveled on one edge
Curved center support
1x nailer
2x6 hip rafter
2x2 spacer
Plywood gusset
Fascia

In this roof, I installed Filon skylights to help brighten the inside of the building. Filon (Sequentia Corp., P. O. Box 360530, Strongsville, Ohio 44136; 216-238-2400) is a translucent, fiberglass-reinforced panel that is installed flush with the roof. I made sure that the roof panels overlapped the edges of the skylights by 6 in. and capped each with three ½-in. by 1½-in. redwood battens to keep the panels from flexing during high winds. The battens should extend the life span of the Filons.

Installing hip flashing—Hip flashing was made from light-gauge 16-in. wide aluminum bent down the center. At the lower edge of each hip, where corners sweep upward, the cap pieces needed to conform with the roof curves. I used 1½-in. steel pipe, flaring the edges of the hip flashing over the pipe, to form a uniform radius free of dents. Pieces of flashing, 8 in. and 12 in. long, were overlapped and fitted to the curve, nailed and then pop-riveted along the edges to keep them flat. After fit and installation, the outer flaps of each corner piece were folded under the edge of the roofing for support and a smooth termination.

I let the roof weather for a month to let the rain wash off a thin manufacturing film from the aluminum. Then I washed, primed and painted the panels. The roof is used to collect domestic water, so I chose a potable-grade of latex paint as the final coating. □

Tim Carr is a furnituremaker and architectural designer in Haiku, on the island of Maui, in Hawaii. Photo by Rob Ratkowski.

Framing a Dutch Roof

Hang a hip roof on a gable end for a dramatic roof form

by Larry Haun

Components of a Dutch roof

A Dutch roof combines a gable roof and a hip roof. The gable is framed first, with the first set of gable-common rafters placed a set distance from the end of the building. The hip roof is fastened to the first gable rafters, which are reinforced to support both the hip rafters and the Dutch ridge.

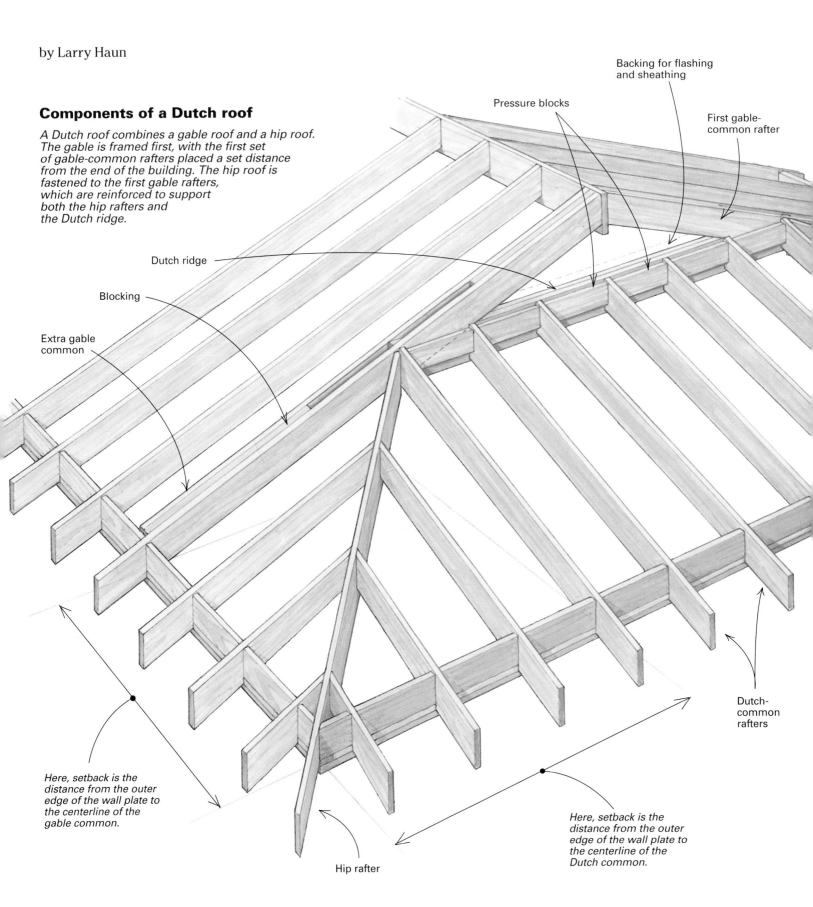

Backing for flashing and sheathing

Pressure blocks

First gable-common rafter

Dutch ridge

Blocking

Extra gable common

Here, setback is the distance from the outer edge of the wall plate to the centerline of the gable common.

Here, setback is the distance from the outer edge of the wall plate to the centerline of the Dutch common.

Dutch-common rafters

Hip rafter

Drawing: Bob Goodfellow

I recall the first time I was given house plans with a Dutch roof (sometimes called a Dutch hip or Dutch gable). The plans looked difficult even though I already knew how to frame both a gable and a hip roof. When I realized that the Dutch was really just a combination of the roofs I already knew how to frame, the plans looked simple, and the roof went together fairly easily. Since then, I've framed plenty of Dutch roofs.

Build the gable first—The Dutch is a section of hip roof in which hip rafters run into a gable end instead of going up to a ridge (drawing below). Exactly where the gable interrupts the hip is called the setback, the distance from the end of the building to the first set of gable-common rafters. When stick framing a Dutch roof, I begin by checking the plans for the amount of setback. A long setback means that the roof will show more hip than gable; a short setback shows more gable than hip.

On the house we featured here, the plans specified a 5-in-12 slope on the gable roof with setback of 5 ft. 6 in. to the center of the first pair of gable-common rafters. I lay out this distance on the top plates and cut and build the gable section of the roof (see p. 25).

Use rafter tables to lay out commons, hips and jacks at the same time—The hip section of the Dutch roof requires several framing members: hip rafters that run 45° from the corners,

Dutch-common rafters that run perpendicular to the gable end and hip-jack rafters that die into the hip rafters. To figure out the lengths of all of these framing members, I use the setback distance, 5 ft. 6 in., and the roof pitch, 5-in-12, and I consult a rafter table (*The Full Length Roof Framer*, A. F. Riechers, Box 405, Palo Alto, Calif. 94302). You also could use a ft./in. calculator (Calculated Industries Inc., 4840 Hytech Drive, Carson City, Nev. 89706; 800-854-8075).

The 5-ft. 6-in. setback is also the run—the horizontal distance covered by the rafter as seen in the plan view—of the Dutch-common rafters. Doubling this figure gives me the span, the dimension necessary for using most rafter tables. In this case the span is 11 ft., and the rafter table's common-rafter column shows the length of the Dutch-common rafters for this roof as 5 ft. 11½ in. to the center of the Dutch ridge. I subtract half the ridge thickness, ¾ in., to find the exact plate-to-ridge length of the Dutch-common rafters; then I mark and cut them.

Jack rafters are laid out in pairs and cut with a bird's mouth on one end just like the commons, but they have a 45° cheek cut (photo, bottom right) where they butt into the hip rafter. The first pair of jacks is shorter than the common rafter, and each successive pair is shorter than the previous pair. The difference in length is the common difference and can be determined using the rafter table. For a 5-in-12 pitch, the common difference for jacks spaced 16-in. o. c. is 1 ft. 5⅜ in.

Scribing and cutting hip rafters—The lengths of hip rafters are listed under the hip/valley column in the rafter table. For a 5-in-12 pitch, the hip for an 11-ft. span is 8 ft. 1¼ in. This is measured from the ridge (a Dutch ridge in this case) to the plumb cut at the end of the bird's mouth.

Just like common rafters, hips have to be shortened by half the thickness of the ridge. Unlike commons, which meet the ridge at a right angle, hips come in at 45°, so they have to be shortened by half the 45° thickness of the ridge, or 1¹⁄₁₆ in. This dimension yields an adjusted hip-rafter length of 8 ft. ³⁄₁₆ in.

A Dutch roof has two hips, so I place two pieces of hip stock, crowns up, on the horses. Hip stock should be 2 in. wider than the commons and long enough to include the tail that forms the overhang. The plans call for an 18-in. overhang; the actual length of the hip tail, about 26 in., comes from the rafter table.

I mark the ridge plumb-cut location square across the top edges of the hip stock, then scribe

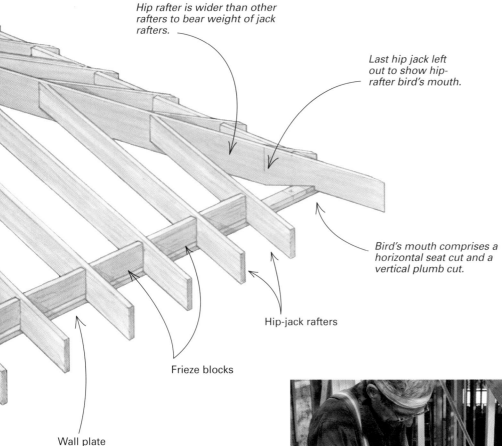

Hip rafter is wider than other rafters to bear weight of jack rafters.

Last hip jack left out to show hip-rafter bird's mouth.

Bird's mouth comprises a horizontal seat cut and a vertical plumb cut.

Hip-jack rafters

Frieze blocks

Wall plate

Rafter templates speed layout. In the small photo at far left, the author positions a hip-rafter template to mark the bird's mouth. Hips are wider than commons and jacks, so the left half of the template is used to scribe the hip tail, which must be ripped down to the width of the jacks and the commons. In the near-left photo, the author uses a different template to scribe a cheek cut on a jack rafter.

Use the Dutch commons to locate the Dutch ridge. Using a Dutch-common rafter as a template, the author marks its plumb cut on the first gable common. After the opposing gable common is marked, the Dutch ridge will sit even with the top of these lines.

the plumb lines with a hip-rafter template. (A rafter template is like a short rafter in that it's got the ridge cut, the bird's mouth and the rafter tail all on a piece of 1x the same width as the rafter but only about 2-ft. long. I make templates for each type of rafter in a roof.) The ridge plumb cut can be a 45° cheek cut or a double-side cut: two 45° plumb cuts that form a point. I've never been convinced that one cut is better than the other, but on this roof I made a double-side cut just for the sheer enjoyment. To make this cut, I use the tongue of the framing square to mark a second plumb-cut line 1½ in. away from the first one. With the saw set at 45°, I make the first side cut; then I go on to make the second cut in the opposite direction.

Next, I flush up the ridge ends of the hip stock, measure out 8 ft. ³⁄₁₆ in. (the adjusted rafter length) and mark for the heel plumb cut on the top edge of the rafters. I align the hip template's registration mark with the mark for the heel cut

on the hip stock and scribe the bird's mouth, which, because this is a hip rafter, has a deeper seat cut than a common rafter (bottom left photo, p. 69). Then I use the template to scribe the tail part of the hip rafter and rip it down to the width of the common and jack rafters.

The Dutch ridge is scribed in place—The next step is to use a Dutch-common rafter to mark the location of the Dutch ridge that hangs between the first pair of gable commons. I place the Dutch common flush alongside each gable common, seat cut to seat cut, and scribe the plumb cut (photo above).

Next, I pick out a 2x that's 2 in. wider than the Dutch commons. This 2x will be the Dutch ridge, which supports the top ends of the Dutch commons. With a little help, I hold the ridge stock in place against the gable commons so that it crosses them at the top of the plumb lines I've just scribed. I scribe each end of the ridge by mark-

ing on the underside of the gable common; then I cut the ridge to length. Next, I spike the ridge in place up under the commons with several 16d nails (top photo, facing page). Later, I'll reinforce it with a backer.

Blocking separates and strengthens the rafters—Now, I pull up the hip rafters, position them directly over the outside corners and fasten them to the wall plate with two 16d toenails on one side and one on the other. I nail off rafters at the plate first with the hip rafter centered on the corner. At the ridge, the plumb cut lies flat against the gable commons. I drive three 16ds through the commons into the hips.

The 5-ft. 6-in. setback means that the first set of Dutch commons is positioned 5-ft. 6-in. o. c. from the corners of the building. Each common is nailed to the top plate with two 16d toenails on one side and one 16d on the other. At the ridge, the first Dutch commons butt against the side

The Dutch ridge is nailed between the first pair of gable-common rafters. After striking a chalkline to position the top of the Dutch ridge, the author and his brother nail the 2x10 ridge between and under the 2x8 gable commons. The Dutch-common rafters hang from this Dutch ridge.

Blocking prevents twisted rafters. The author nails a 2x pressure block to the Dutch ridge. A pressure block goes between all of the Dutch commons.

2x6 backing strengthens the Dutch ridge. Because the Dutch ridge only was nailed under the gable commons, a backing ridge is face nailed to the gable commons and to the Dutch ridge for added support.

Double the first gable for support. The gable end supports the hip roof, so a second set of gable commons is installed. The gap is filled with 2x blocking.

cuts of the hip rafters, and they're spiked in place with two 16ds.

Next, I nail with three 16ds a 14½-in. pressure block, a 2x block the same size as the common, to the Dutch ridge and tight against the common (bottom left photo, above). The front edge of the pressure block is flush with the common. The pressure block helps keep the rafter in place.

At the plate line a 14½-in. 2x frieze block is nailed between each rafter with one 16d at one end and two at the opposite end. This block helps keep the rafter from rolling over and strengthens the roof structure. Then I install another common and another pair of blocks and so on until I get to the jack rafters.

Before nailing in the hip-jack rafters, I sight down the hip rafter and make sure it's straight from the ridge to the plate. If the hip is bowed, I temporarily brace it straight until the jacks are nailed in place. Then, beginning with the longest jack, I nail it at the plate against a 14½-in. frieze

block, driving two 16ds into the block and one 16d toenail into the plate on each side of the jack. I nail each jack to the hip with three 16d nails, taking care not to bow the hip rafter from side to side. Once the opposing jack is nailed in, the hip is locked in place. Then I install the rest of the jacks and frieze blocks.

Once all the jacks and commons are nailed in, the overhangs can be measured, marked and trimmed to length, and the fascia can be nailed on. The length of the hip overhang is determined simply by extending the chalkline on the commons all the way across the hip. When cutting rafter tails to length, I check to be sure I'm using the common-rafter template on the commons and the hip-rafter template on the hips.

The Dutch ridge needs extra support—With all the rafters in place, it's time to reinforce the ridge and commons holding the upper part of the Dutch roof. I nail a long 2x, 2 in. wider than

the Dutch ridge, to both the Dutch ridge and the two supporting gable commons (bottom center photo, above). The ends of this backing ridge are flush with the top of the gable commons.

Next, I place another set of gable commons against the backing ridge, space them from the two ridge-supporting gable commons with 2x blocking and nail the new set of gable commons to the plate, the gable ridge and the backing ridge (bottom right photo, above). This second set of gable commons and the backing ridge provide plenty of support for the Dutch roof. The last step is to hold a 2x6 directly over the Dutch ridge, mark, cut and nail it to the backing ridge. This 2x6 acts as backing for flashing needed between roofing materials and siding. □

Larry Haun of Los Angeles, California, is a framer and author of The Very Efficient Carpenter, *a book and video series published by The Taunton Press. Photos by Scott Gibson.*

Putting the Lid On

A primer on production cutting and raising a hip and gable roof

by Don Dunkley

One of the most satisfying events in building a house is the completion of the roof. Some builders borrow from European tradition and nail a pine tree to the peak in celebration. At the least, it is usually the excuse for a party. There are good reasons to celebrate. Framing a roof can be perplexing, physically taxing and sometimes dangerous. However, with thoughtful organization of rafter layout, production rafter-cutting techniques and carefully built scaffolding and bracing to help raise the ridge and rafters, your celebrating doesn't have to come out of a sense of relief.

The best way that I know to share my knowledge of roof framing is to describe the steps involved in building a simple hip and gable roof, like the model roof that is shown in plan, below. This article will cover most of the problems that are encountered in a rectangular building—laying out and assembling common rafters, hips and jacks, along with the ridge, purlins and collar ties.

Preparation—The roof is ready to frame once all the walls are built, plumbed up and braced off. The exterior walls must be lined very straight, because any irregularities in the span will show up on the roof frame. Before you start sorting through your framing stock, study your roof plans carefully. They should show an overhead (plan) view on a scale of ⅛ in. or ¼ in. to 1 ft. They will tell you the type of roof (gable, hip or gambrel), the pitch or slope, the length of overhangs (eave and gable end), the layout of the rafters, their spacing (16 in. on center, 24 in. o.c.), and the sizes of the framing members.

Layout—Job-site layout begins with measuring the span of the building. Always measure from the top (double) plate height. There are usually slight variations between the span shown on the plans, the actual span at the bottom-plate level, and the one at the double plate. Since rafter lengths are calculated down to ¼-in. changes in span, use the double-plate measurement. A 100-ft. tape is the tool for this job.

First, as shown in the photo below, the positions of the rafters must be marked on the top of the double plate. This lets you properly locate the rafters when erecting the ridge. The layout is also necessary to distinguish the positions of the rafters from those of the ceiling-joist layout, which should be placed so they can be used as ties to which the rafters can be nailed. Starting with the hipped end of the roof, lay out the positions of the three king common rafters. Strike a line 10 ft. in from each corner down the length of the building, as well as one midway along the width, and write the letter C (for center) on the plate over each of these lines, which will serve as centers for the king commons. Next, lay out hip-jack rafters on 2-ft. centers from the corner of the building toward the king common rafters.

The common rafters are laid out similarly on the plates, starting at the gable end. I usually mark one side of the rafter position with a line across the top plate. If ceiling joists are also on 2-ft. centers, you don't need to lay them out, because they will be installed beside the rafters. If joists are on 16-in. centers, you would start the layout with the tape held 1½ in. past the end of the top plate. This way, a joist will tie into a rafter every 4 ft.

The ceiling joists that sit on the exterior wall will stick up above the rafters, and can be trimmed along the pitch of the roof after the rafters are up, and before the decking is applied. On the hip, the ceiling joists close to the end wall can't be nailed in place unless you notch them or cheat them off the layout, because the hip will interfere. They should be laid flat on their layouts and installed after the hips and jacks are in place.

Layout tees—The layout tee is a handy tool that lets the builder lay out rafters accurately and quickly. It also helps eliminate steps in rafter-length calculations. Layout tees should be made for the bird's mouth and tail of both

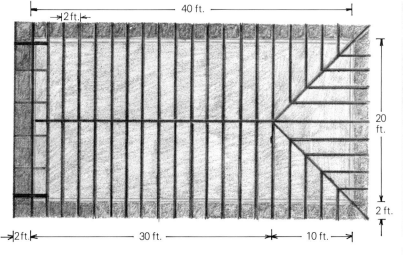

The roof plan of the model, above, shows a gable end using a barge rafter and outriggers for a 2-ft. rake overhang, and a hipped end with a 2-ft. eave. The 2x6 rafters are on 24-in. centers, and the roof pitch is 8-in-12. The span in this case is 20 ft. Right, a carpenter lays out the joists and rafters by walking the plate, something that should be done only after the walls have been plumbed, lined and well braced.

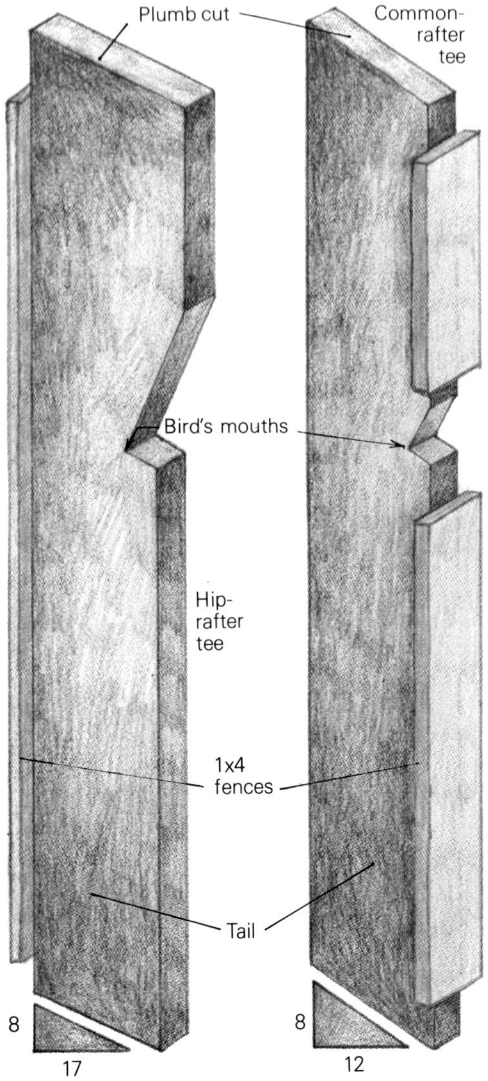

common and hip rafters (drawing, above). There should also be a plumb cut at the top of the tee above the bird's mouth to use as a pattern for the top plumb cut.

Tees should be made of the same width stock as the rafters, so in this case the layout tee for the common rafter should be made from a 3-ft. scrap of 2x6. After you scribe the plumb cut at the top of the tee, move the square about 12 in. down from the top plumb mark and scribe out a bird's mouth. Next, mark the tail cut by measuring along the body of the square for the length of the overhang, which is 2 ft. in this case. Then make a mark and scribe the plumb tail cut, for an 8-in-12 pitch in our example. Cut out the pattern and nail two pieces of 1x4 along the bottom of the tee, one staying clear of the bird's mouth and the other not projecting past the top plumb cut, as shown above. When the tee is being used, the 1x4 fence registers against the bottom edge of the rafter stock for marking out the bird's mouth and plumb cuts.

The hip-rafter tee in this example is cut from a 3-ft. 2x8. On one end, scribe an 8 and 17 plumb mark. Move the square down about 12 in. and scribe the seat cut. Since this is a hip, the seat cut must be dropped (cut more deeply) to bring the top edge of the hip into the same plane as the jacks. To determine the amount of drop, lay the square at the top of the 2x8 on an 8 and 17. Where the 17 mark intersects the top of the lumber, measure down the body of the square half the thickness of the hip (here ¾ in.) and make a mark. Then measure down from the top of the lumber, perpendicular to the edge, to this mark (5⁄16 in.). This is the drop needed. Make the hip seat cut 5⁄16 in. deeper than the common seat cuts.

The width of the tail of a hip must equal the width of the common rafter, so the wood past the seat cut must be ripped down from a 2x8 to a 2x6. Measure down from the top edge of the rafter 5½ in. (the actual dimension of a 2x6), mark the length of the rafter tail, and rip the excess 2 in. off the rafter's bottom edge. Since this rip creates a step in the bottom edge, both pieces of 1x4 fence should be nailed to the top of the tee. When using this tee, place it on the top edge of the rafter stock.

The next job is working up a cut list for all the rafters. Count the rafters on your plans, and calculate their lengths. The cuts can then be scribed using the rafter tees, and all the pieces can be cut before beginning the actual installation. This approach requires both confidence and intense concentration, but doing all the cutting first speeds up the process by letting you put your head down and frame without having to stop and figure.

The rafter book—I wouldn't want to be without a rafter book when framing roofs. Mine contains 230,400 rafter lengths for 48 pitches. I can look up any building span under the appropriate pitch, and quickly determine the rafter length and angle of cut. This book saves a lot of labor, and eliminates many costly errors.

Calculating common-rafter lengths—In the rafter book under 8 in 12, the common-rafter table shows that our span of 20 ft. requires a rafter length of 12 ft. ¼ in. from heel cut to plumb cut. This measurement doesn't account for the ridge reduction, because ridge thickness is not a constant. With a 2x ridgeboard, the reduction along a level line is ¾ in. But measured along the rafter edge, ¾ in. measures ⅞ in. on an 8-in-12. Rather than laying out a shortening line on the stock, I subtract the ridge reduction measurement from the rafter-book length to get the corrected rafter length down to the bird's mouth—in this case, 11 ft. 11⅜ in.

The overhang length from the heel cut to the tail cut can be taken off the rafter tee or determined from the rafter book by adding the overhang for each run to the span. A 2-ft. overhang will add 4 ft. to the span. In the rafter book, the 24-ft. span at 8 in 12 reads 14 ft. 5⅛ in. Deduct from that figure the full rafter length of 12 ft. ¼ in. This leaves 2 ft. 4⅞ in. in length from the heel cut to the toe of the tail cut. The overall length of the rafter will be 14 ft. 4¼ in.

Calculating lengths of hips and jacks—Use a large pad of paper to organize your calculations for the hips and their jacks, since they involve a bit of figuring. On the job, keep your building plans clean. Don't scribble math all over them. Using the rafter book, an 8-in-12 hip at a span of 20 ft. is 15 ft. 7⅝ in. A ridge reduction is necessary, and this 45° thickness measures 1³⁄16 in. along the rafter edge at 8-in-12. This reduces the rafter length to 15 ft. 6⁷⁄16 in. from plumb cut to heel cut.

To find the overhang or tail length, add 4 ft. to the span, just as for the common. The rafter book lists 18 ft. 9⅛ in. for a 24-ft. span. This leaves a rafter tail of 3 ft. 2½ in., and an overall length of 18 ft. 8¹⁵⁄16 in.

To calculate the lengths of the hip jacks, look up jack rafters on 2-ft. centers at 8-in-12. Both the square and the rafter book read 2 ft. 4⅞ in. This distance is the common difference, or how much longer one jack will be than the previous one. This is also the length of the first jack before the deductions. If you are using my system of subtracting the ridge reduction (measured along the edge of the rafter) from the rafter-book length, then subtract 1⅜ in. on an 8-in-12 jack to get 2 ft. 3½ in. from plumb cut to heel cut. Only the first jack needs to be figured for the deduction since the rest will automatically follow, as the common difference is added to each one.

Cutting the rafters—With all the calculations complete, the next step is to lay out and cut the rafters. You can use production techniques that save a lot of time without sacrificing accuracy. I use a rafter bench, an oversize, site-built sawhorse that holds ganged rafter stock up off the ground for easy marking and cutting. I try to set up my benches close to the lumber stack, which should be fairly close to the building. You will be worn out before you start if you have to carry a ton of rafters a great distance.

Stack all the rafters of one type on the bench with their crowns down. The crown is a convex edge seen by sighting down the lumber. Crowns should be placed up in construction to help deflect the load placed on the rafters or joists; they are stacked crown down on the rafter bench so you can scribe cut-lines with the

A chalk line snapped across ganged common rafters marks the heel of the plumb-cut line. As indicated by the layout tee (bottom left) the rafters are stacked with their bottom edges up, and their ends even and square. The layout tee will be used on each rafter to scribe the plumb cut, bird's mouth, and tail cut.

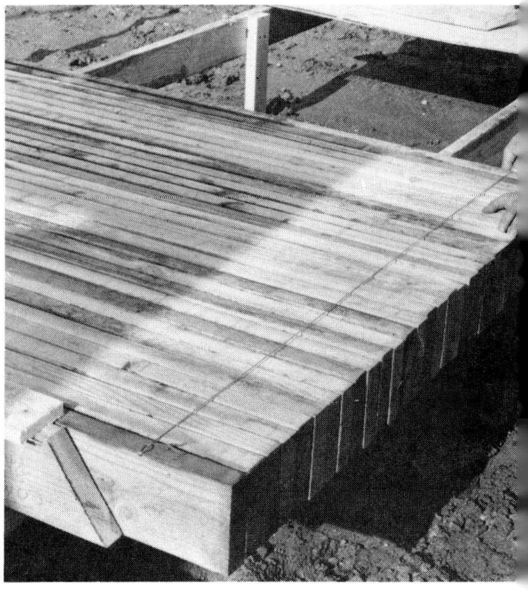

Illustrations: Frances Boynton

Jack rafters stacked on a rafter bench show the common difference of 2 ft. 4⅞ in. on an 8-in.-12 pitch using a 24-in. spacing. The diagonal lines indicate the direction of the side cut that will produce pairs of jacks (left and right) for each hip rafter. The bird's mouth and tail will be marked with the common-rafter layout tee.

layout tees. When you stack the rafters on the bench, keep their ends flush so they can be squared up easily with a framing square by drawing a line across their edges. Then use the layout tees to mark the plumb cut at the top, and bird's mouth and tail cuts at the bottom on each of the outside rafters of the stack. Measure the length you have calculated between the plumb cut and the bird's mouth several times, and then connect the marks across the stack with a chalk line (photo facing page, bottom). Lay the first outside rafter down flat on the bench, scribe the plumb, seat and tail cuts with the rafter pattern, and cut them out.

When all the common rafters are cut, they should be dispersed along the exterior walls according to the layout. Before spreading the rafters out, it is a good idea to set a 16d nail at the top plumb cut. This toenail will come in very handy during assembly.

Cut hip rafters using the same procedure, but make double side cuts at the top (a vertical 45° bevel on each side for the plumb cut). These cuts can be made easily on 2x stock with a circular saw set at 45°. For larger timbers or glue-lams, the angle of the top edge of the stock must be laid out, and the cut made with a handsaw.

With the double side cut complete, measure down from the top of the rafter the distance calculated, 15 ft. 6⁷⁄₁₆ in. Mark this length on the center of the rafter's top edge. Slide the tee to this point on the rafter, and scribe the seat cut and heel cut. Mark the rest of the board, scribing along the tail of the pattern. Rip the tail down to the proper width.

Use the common-rafter tee for the jack layout. Group the jacks on the bench according to length—for the model roof, there will be four sets of four. Load the longest first and work down to the shortest set. Only the tail ends can be squared up. Lay the common-rafter tee on this end and scribe the tail and seat cuts on the outside rafter. Then lay out the rafter on the opposite side of the stack and snap lines.

To lay out the plumb cut at the top of the jack, measure the length of common difference—2 ft. 3½ in.—up from the seat-cut line for the shortest set, and add 2 ft. 4⅞ in. progressively to each set of jacks. Square these marks on the top edge of the rafters, and lightly mark two of each set with a 45° line indicating the direction of the angle. Mark the other two with the opposite angle (photo left). The side cut must be laid out this way because the length of a jack is measured from its centerline. Scribe a 45° line in the direction of the light line drawn previously through the center of the plumb-cut line on the top edge of the rafter. Then place the layout tee at the end of the 45° line that intersects the edge of the board farthest down the rafter, and scribe the plumb cut on the face of the rafter. This method creates a slight inaccuracy in the length of the jack on a moderately pitched roof, but it is much faster than marking the precise angle (which can be found in the rafter book or on the square) on the top edge of each rafter.

After cutting all the jack pairs, set them on the roof, paying particular attention to the correct placement of right and left-hand rafters. Drive a 16d nail into the smaller jacks, and hang the head and shank of the nail over the double plate so that the jacks hang along the wall, out of the way but still accessible.

The last roof members to get cut are the ridgeboards and purlins. The 30-ft. ridgeboard on the example here is made from two pieces. Pick straight stock, and cut so the break falls in the middle of a common-rafter layout. The board that includes the hip end of the building should be left long by 6 in., and all cuts should be square.

Assembling the roof—The reward for all the calculating, laying out and cutting is a roof whose members fall right into place once the ridge is up. This is the stage with the largest element of danger, and safety is a primary concern. While nailing joists and laying out the top plate, you'll start to develop "sea legs," gaining confidence in walking around up there. Make sure no loose boards stick out more than a few inches beyond a joist, and keep the top plates between joists free of scraps and nails.

Establishing the ridge height—Before doing anything else, calculate the ridge height to see if you need scaffolding to install it. This is done by multiplying the unit of rise (8 in our example) by the run of the building (10). The bottom of this ridge is 80 in. from the top plate. Ridges 6 ft. or more above the plate need scaffolds. A good scaffold is about 4 ft. lower than the ridge. It must be sturdy, well braced, spanned with sound planks, and running down the center of the building. To make room for the placement of ridge supports and sway bracing, leave a 1-ft. wide space between the scaffold planks.

Raising the gable end and ridge—First, put the tools and materials where you need them. Saws, nails and other tools can be kept handy on a sheet of plywood tacked on the joists. Pull the ridgeboards up on the joists alongside the

scaffolding. You'll need several long 2x4s for braces and legs. Stack them neatly on the joists along with 2x8 bracing for the purlins. To support the gable-end rafters in the initial stage of assembly, nail two uprights to the gable-end wall, perpendicular to the top plate and a foot on each side of the center.

For setting the gable and ridge, you need a crew of four—two carpenters on the scaffold, and one at each end of the span. Starting at the first rafter on the gable end, the carpenters on the outside walls pull up the gable-end rafters, setting the top plumb cut on the scaffold. A small 2x4 block 7½ in. long, the height of the ridgeboard, should be nailed to the plumb cut of one of the rafters. This block temporarily takes the place of the ridgeboard. Make sure the block is flush with the top of the plumb cut. The carpenters on the scaffold pull up the rafters until the seat cuts sit flush on the top plate. The carpenters on the outside walls nail the rafters down, keeping the outside of the gable-end rafter flush with the outside wall. Toenail each rafter to the double plate with two 16d nails on one side and one 16d on the other (back nail). At the plumb-cut end, the rafter with the temporary block must align with the other rafter so that the cuts are nice and tight to the block.

When the gable-end rafters are in position, nail each rafter to the uprights with 16d nails. You'll need to insert a temporary support under the ridge. Measure down from the bottom of the block to the top plate on the gable-end wall to find its length (drawing, below). Nail the leg down to the plate where the 10-ft. center is marked. You'll need another leg under the joint in the ridgeboard, but before you cut it, look for something to set it on. If there isn't a wall directly below the ridge, lay a 2x6 or 2x8 across the joists to carry the leg. After this leg is cut, the block can be removed from the gable end and replaced with the ridgeboard. The carpenter on the other end of the ridgeboard should rest it on the support leg, and scab an 18-in. 2x4 onto the leg and ridge. The scab should stop at least 1 in. below the top of the ridgeboard.

After one end of the ridge is raised, install the common rafter pair that is one layout back from the other end of the first length of the ridgeboard. When you're nailing rafters to the ridge, use three 16d nails to face-nail the first

Supporting the gable end and ridge

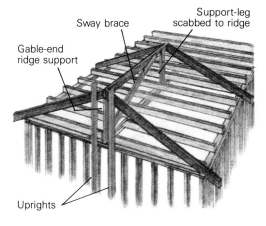

Sway brace

Support-leg scabbed to ridge

Gable-end ridge support

Uprights

The king-common rafter that butts the end of the ridgeboard is try-fitted and used to scribe a line for cutting the ridgeboard in place (left). The rafter in the foreground is a king-common that will be nailed at the end of the ridgeboard, perpendicular to the rafter being used for scribing. The skeleton formed by the three king-commons (center left) supports the ridgeboard so the common rafters can be nailed up with frieze blocks. The vertical 2x8 in the foreground is a temporary gable-end brace.

Bottom left: The underside of a hip rafter shows the jack-rafter pairs in position. The upright brace under the hip is placed over a wall. Hips are cut from stock 2 in. wider than common rafters to accommodate the width of jacks cut at compound angles. The added width gives strength for the long span required of hips.

rafter in the pair; then toenail the second. When these rafters are secured, the gable end should be plumbed, and temporarily secured with a swaybrace, a 2x4 with one end cut on a 45° angle, that reaches from the plate (or a 2x8 nailed to the joists above) to the ridge.

Installing hips—The remaining length of ridgeboard is set next. This is easily done by nailing another support leg at the end of the new ridge piece and setting the two king common rafters that define the hip. The third king common, the one that nails to the end of the ridge, is next (photo top left). The hip end of the ridge should extend about 6 in. beyond the king-common layout to allow for final fitting. Do not nail the third common yet, but slide it up to the ridge and scribe the ridge at the plumb cut when the rafter is flush to the top of the ridge and seat cut is up tight (photo center left). Set the rafter down and cut the ridge off, then nail it in place. The resulting frame should be plumb and strong, and ready for the hips.

Raise the hip, pushing its double side cut into the slot at the ridge, and toenail it at the corner and at the ridge. If the hip is spliced, haul the pieces separately on the roof and nail a 2x4 cleat to the bottom edge of the hip at the scarf joint. Position it on the bottom edge so it doesn't interfere with the jacks. Pull a string from the top center of the hip at the ridge, down to the center of the hip at the seat cut. Nail in a temporary upright under the center of the hip and align it with the string. This should eliminate any sag. If it is spliced, cut a leg to fit under the cleat (photo bottom left). Now you can nail the jacks and their frieze blocks.

Jacks, commons and frieze blocks—Start with the smallest jacks and work up. Nail in pairs, to avoid bowing the hip. Then nail the seat cut.

The remaining common rafters can now be filled in, followed by the frieze blocks that go between the rafters at the double plate. The blocks for a 2-ft. o.c. spacing should be cut 22⁷⁄₁₆ in. and driven tight. Frieze blocks that fit against the hip will have a side cut on one end (photo right). Frieze blocks that are to be nailed perpendicular to the rafters should remain full height. However, if they are to be nailed plumb, they will have to be beveled on the

pitch. This is most easily done on the table saw, but you can do it with a skill saw. In either case, use rafter off-cuts and discarded rafter stock for frieze blocks. For repeated crosscuts, use a radial arm saw; alternatively, you could set up a simple cut-off fixture for your circular saw. The blocks for our example are held to the outside of the top plate, square with the rafter (not plumb) and toenailed flush at the top of the rafter. The next rafter on layout is then pulled up, set in place, nailed at the seat cut and the ridge, and then nailed through the side into the frieze block behind it. Drive two 16d nails for 2x6s; three 16d nails for 2x8s. Whenever a ceiling joist lands next to the rafter, drive three 16d nails through the rafter into the joist.

Purlins—Purlins are required where rafter spans are long. Purlins run the length of the building at the center of the rafter span. They are usually made of the same stock as the ridge, and should be positioned once all the rafters have all been nailed in place. If the commons are 18 ft. or over, it's much easier to handle them if the purlins are installed beforehand. To put up the purlin, first string a dry line across the path of the common rafters to check their sag at the center of their span. Start a purlin at one end, and toenail it into the bottom edge of the rafter, while it's being held by two carpenters. It is held square to the edge of the rafter and perpendicular to the rafter slope. Toenail it to the rafters in several places. Then cut legs (kickers) to fit under the purlin (small photo, facing page). The kickers must sit on the top of a wall, and to avoid deflection should not be placed in the middle of a ceiling-joist span.

Finishing up—Gable roofs are also reinforced with collar ties—horizontal members that connect one rafter in a pair to its opposing member. Collar ties should be no lower than the top one-third of the rafter span. Measure down from the ridge along the slope of the rafter and make a mark about one-third of the way down. Now mark the same distance on the opposite rafter. Hold a 2x4 (or wider board) long enough

This hip rafter has been toenailed in place. The frieze blocks required a single side cut for their intersection with the hip. In cutting the bird's mouth for the hip, the amount of drop had to be calculated. This meant taking a deeper cut so that the top edge of the hip is in the same plane as the other rafters.

The purlin in the foreground (above) is supporting the span of common and hip-jack rafters. Braces positioned at interior walls are perpendicular to the slope of the roof.

Standing on the outriggers (right), a carpenter nails the barge rafter. The frame has been notched for the flat 2x4 outriggers, which are face-nailed to the first rafter inside the gable end, and flat-nailed to the gable-end rafter. The rafters are the top chords of Fink trusses.

to span the two rafters at the marks, and scribe it where it projects past the top of the rafters. Using this as a pattern, cut as many collar ties as you need.

The gable ends must be filled in with gable studs placed 16 in. o.c. Each gable stud fits flush from the outside wall to the underside of the gable rafter. You can make the gable stud fit neatly under the rafter by making square cuts with your saw set on the degree that corresponds with the pitch of the roof. For an 8-in-12 pitch, the corresponding angle is $33\frac{3}{4}°$. You can find the degrees in the rafter book under the pitch of the roof. Gable studs are best cut in sets and, like jack rafters, they advance by a common difference.

The example shows a rake of 2 ft., with a barge rafter and outriggers. Unlike the fly rafter and ladder system shown in the glossary (see p. 140), a barge rafter usually isn't reduced for the ridge; it butts its mate directly in front of the end of the ridge board. The outriggers support the barge overhang. They are typically 2x4s, 4 ft. o.c. from the ridge down, extending from the barge rafter across the gable-end rafter and beyond one rafter bay. The outriggers are notched into the gable rafter, laid in flat and face-nailed to the common rafter in back, as shown in the photo at right.

To put in outriggers, first lay out the top of the gable rafter 4 ft. o.c., starting from the ridge. The layout should be for flat 2x4s ($3\frac{1}{2}$ in. wide). Then notch the layout marks with several $1\frac{1}{2}$-in. deep saw kerfs and a few quick blows from your hammer. Make these cuts down on the rafter bench. Let the outriggers run long and cut them along a chalked line once they are up to ensure a straight line for nailing the barge rafter. □

Don Dunkley is a carpenter and contractor in Sacramento, Calif.

No valley rafters. When framing two roofs that intersect, the common method is to use valley rafters (drawing below left). A faster method (drawing below right) is to sheath the main roof first, then frame the smaller roof on top of it (photo above).

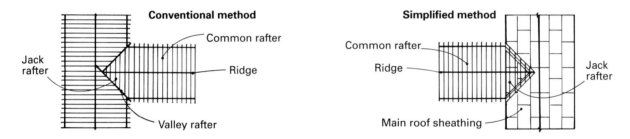

Conventional method

Jack rafter — Common rafter — Ridge — Valley rafter

Simplified method

Common rafter — Ridge — Main roof sheathing — Jack rafter

Simplified Valley Framing
Build one roof on top of the other and skip the valley rafters

by Larry Haun

The foreman on one of my first framing jobs asked me if I knew how to build a California roof. I had to admit I didn't. Instead of firing me on the spot, the foreman gave me until the next morning to learn. I pored over my carpentry books at home that night, and I found what I was looking for under "blind valley." That after-hours discovery nearly 40 years ago helped me keep my job, and the framing technique I learned is just as useful to me today.

When framing two gable roofs that meet at right angles, and when one of the roofs has a lower ridge than the other, the common approach is to use a supporting valley rafter that extends

from the wall plate to the main roof ridge, and a shorter nonsupporting valley rafter that intersects it (see the article on pp. 20-24). But jack rafters must then be cut to fill the triangular space between the ridge and the valleys on the main roof (left drawing, above). And a second set of jack rafters is needed for the smaller roof. All of that takes time.

The blind valley, or California roof as it's called out here, often is a less-complicated way of handling the same situation (right drawing, above). The technique certainly makes sense when framing an addition because the new roof can be framed directly on the old roof. It also works well

in new construction when the room under the main roof has a cathedral ceiling. Although this is most often used when a smaller roof intersects a main roof, the same technique can be used when the ridge heights of the two intersecting roofs are the same.

I was stumped by this framing problem as a green carpenter back in the 1950s, but the technique is not difficult. The main roof is framed, and the roof sheathing is applied. Then the common rafters of the smaller, intersecting roof are erected, and the ridge is carried over to the main roof. Finally, the valley jack rafters of the smaller roof are cut and installed, linking

Drawings: Mike Hiotakis

the ridge of the smaller roof with the deck of the main roof. The photos in this article show this technique being used to frame a new roof on an existing roof covered with a membrane.

Some building codes require full sheathing under the secondary roof to maintain the main roof's shear strength. Full sheathing is certainly what you would use when building an addition onto an existing roof. At minimum, the main roof must have sheathing where the secondary roof ridge and jack rafters land. You may need to leave a hole in the sheathing on the main roof to allow passage for people or ductwork

Extending the ridge—Once the main roof has been built and the sheathing applied, the common rafters of the secondary roof are raised. The inboard end of the ridge extends to the main roof if the stock is long enough. If not, the ridge needs to be extended. To find where the ridge should meet the main roof, sight along the length of the ridge and mark the point where the top of the ridge falls on the main roof. A measurement from the mark on the main roof deck to where the secondary ridge ends gives you the length of the ridge extension (top photo, right).

You can use your framing square to mark the angle on the end of the ridge extension that will make the ridge snug on the main roof. The angle will be the same as for the seat cut on the common rafters. But I usually scribe the angle in place (middle photo, right). I set one end of the ridge-extension stock on top of the secondary ridge, and the other end on the main roof at the point where the ridge will end up—the ridge extension should be sitting level at this point. Then I use a scrap of 2x ridge stock to scribe the angle of the main roof onto the ridge extension. Once the ridge extension has been measured and cut, nail it in place on top of the main roof sheathing, making sure it is level and straight. At the other end, toenail the extension to the end of the ridge already in place (bottom photo, right).

Marking valley locations—Now you can mark the location of the valleys on the main roof sheathing. You will need to snap a chalkline from the end of the extended ridge where it meets the main roof to a point near the eaves where the two roof planes come together (top photo, p. 80). To find the lower mark for the valley chalkline, extend the plane of the secondary roof into the main roof—with a stringline or a piece of 2x stock—and find a spot where the two intersect. Snap the chalkline from the ridge, through the spot that you've marked.

You must add support along this line as a base for the tail ends of the valley jack rafters—the short rafters that extend from the valley to the ridge of the secondary roof. Make this base by nailing two 1x6s side by side, or by using strips of plywood 12 in. to 16 in. wide, next to the valley chalkline. Don't nail them on the line. Instead, hold the boards back from the line so that the top edge of the jack rafters will be in the same plane as the line (middle photo, p. 80). The steeper the pitch, the closer to the line the boards will be. To find the exact distance between the chalkline and the 1x6, stretch a line from the

Mark the ridge length. To find the correct length for the ridge extension, measure from the end of the existing ridge to the point you have marked on the main roof deck. Or lay the ridge extension on top of the existing ridge, run the extension all the way to the main roof and then mark the extension where it should be cut.

Scribing the ridge. To mark the correct angle on the ridge extension where it meets the main roof, you can use a scrap of wood and the main roof deck. Set the extension on top of the secondary ridge already in place, make sure it is level and then use a 2x scrap to scribe the angle on the extension.

Nailing it up. Once the ridge extension has the correct angle on one end and has been cut to length, toenail it to the end of the ridge already in place and to the existing roof deck.

Snapping the line. **After the ridge extension is in place, snap a chalkline to mark the location of the valley. The upper point is where the ridge extension meets the main roof. The lower point is found by extending the plane of the secondary roof into the main roof. In this example the wall of the addition is higher than the wall it intersects on the existing structure.**

Installing support material. **The doubled 1x6 supports in the valley are set back from the chalkline. The steeper the roof, the smaller the distance between the support and the line.**

Marking the jack rafters. **The valley jack rafters are marked in pairs, one for each side of the ridge. The angled marks indicate the direction of the bevel (side cut) on the level seat cut. Set your saw to the correct bevel and make sure the blade is angled in the same direction as the slash mark on the edge of the stock.**

ridge of the secondary roof to the chalkline in the valley and push your 1x6 up to the string.

This support material can just be marked in place on the roof and cut roughly to length. Total accuracy isn't required. Nail the support boards in place with 8d nails through the sheathing and into the common rafters of the main roof.

Cutting the jack rafters—The ridge extension can now be filled in with jack rafters. They are laid out in pairs, one jack for each side of the ridge. The first pair of jacks will be shorter than the common rafters, and each successive pair will be shorter as they move along the secondary ridge and up the main roof. The amount that each jack is shortened depends on both the pitch of the main roof and the on-center spacing of the rafters. But each succeeding jack rafter on a roof will be shortened by the same amount, called the common difference. There are several ways to determine this common difference. An easy way is to use a book of rafter tables. Turn to the chart showing the pitch for the main roof (in our example 4-in-12) and find the on-center spacing for the rafters (in this case 16 in. o. c.). Or you can use the table stamped on the blade of your roofing square. The common difference for a 4-in-12 roof is 16⅞ in. So as you move up the valley cutting the jacks, each pair would be 16⅞ in. shorter than the pair before.

To mark the first pair of jack rafters, use a du-

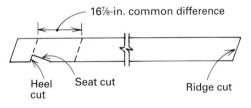

16⅞-in. common difference

Heel cut
Seat cut
Ridge cut

plicate of a common rafter from the secondary roof and lay it on edge next to two pieces of rafter stock that are approximately 1 ft. shorter. Then add two more pieces of rafter stock about a foot shorter than the first pair. Keep adding successively shorter pairs of rafter stock until they shrink to nothing and keep the ends flush at the ridge end. Measure up from the heel cut of the common rafter exactly 16⅞ in. and scribe a mark on the edge of the longest pair. Measure 16⅞ in. from that mark and lay out the next pair and so on until all are laid out.

Each pair of jack rafters will have a plumb cut at the ridge (just like the common rafter) and a level seat cut with a bevel (side cut) on the opposite end to fit the pitch of the main roof. The side cut at the lower end tips either to the right or to the left, depending on which side of the ridge the jack is to be nailed. To indicate this for cutting purposes, make angled slash marks, one to the right and one to the left, on each pair of jacks (bottom photo, left). These marks will help you lay out and cut the correct side cut later.

So far, I have assumed that the last common rafter on the secondary roof falls exactly in the corner. In that case the first jack rafter would be a full 16⅞ in. shorter. If the last common does not fall exactly in the corner, the difference would not be as great, and you'll have to measure to find the length of the first jack. Lay out 16 in. along the ridge and the valley from the last common rafter. Measure from the ridge to the point on the valley chalkline to get the length of the first pair of jack rafters. After that, each pair will be 16⅞ in. shorter than the previous pair.

Making a template—To mark the seat and the ridge cuts on the jack rafters, make a pattern or template (top photo, right). On one end of a short piece of 1x material that's about the same width as the rafter stock, scribe a ridge cut by holding the tongue of the framing square on 4 in. and the blade on 12 in., and mark along the tongue. The level seat cut is drawn on the other end of the scrap by marking along the blade. After cutting along these two lines, nail a 1x2 fence on the top edge to make it easy to keep the template even with the top edge of the stock.

Now, with the template, mark the ridge cuts on all the jack rafters. For the seat cuts, place the template on the jack stock so that the point of the seat cut lines up with the intersection of the length and slash marks you made earlier—this is the long point on the bottom of the jack rafter. Each jack rafter of a pair will get a line for the seat cut—but on opposite sides.

When you're ready to make the seat cuts at the ends of the rafters, your rafter tables will also provide the correct bevel for the side cut. Set your saw to this angle (for a 4-in-12 roof, the bevel is 18½°) and cut along the lines you made with the template. As you cut, make sure the sawblade is angled in the same direction as the slash mark you made on the edge of each jack rafter.

This sounds more difficult than it really is, and it will be obvious when you start doing it. One jack is cut to the left, one to the right. When you're done, you will have pairs of jack rafters that are cut to the same length, but the bevels of the seat cuts will be in opposite directions.

Once the jack rafters are cut, lay out the ridge at the correct spacing (in this example 16 in. o. c.) and nail the jack rafters into place. Use two 16d nails, just like you would for a common rafter. Nail the seat cut to the 1x6 base. Secure these jacks to the base by toenailing them through the sides (middle photo, right) so that no nails will be in the way if you have to cut sheathing.

Finishing touches—If two intersecting roofs have different eave heights—as is the case on the addition I framed for this article—you may need to fit a scrap of plywood or rafter material behind the common rafter where the two roofs meet (bottom photo, right). On a roof with open eaves (i. e., no soffit), this prevents birds from nesting in this corner, and it doesn't leave an unreachable section of roof for the shingler.

If the roofs intersect at the same plate height, you will need to cut a fake valley rafter tail out in the overhang, so the sheathing for both roofs is supported where it intersects. ☐

Larry Haun lives in Los Angeles, Calif., and was a longtime teacher in apprenticeship programs. His book, The Very Efficient Carpenter, *was published by The Taunton Press in 1992. Photos by Larry Hammerness.*

A template will help. To mark rafters accurately, make a template from a piece of 1x material that's the same width as the rafter. Make a ridge cut on one end and a level seat cut on the other. A 1x2 nailed on the top edge of the template acts as a fence to aid alignment.

Installing the jack rafters. Once the valley jack rafters have been measured and cut, they are installed along the ridge just as if they were commons. Toenail them into the 1x6 supports.

When eave heights are unequal. When the intersecting walls are at different heights, a scrap piece of rafter stock or plywood nailed to the backside of the last common on the smaller roof will prevent birds from nesting in the eaves and will simplify the shingling process.

Framing a Second-Story Addition

A quartet of gables is linked by California-style valleys

by Alexander Brennen

Almost every day the local refuse company parks a debris bin in front of some house in Albany, California. The bin usually means that a crowbar-wielding crew will soon arrive to tear the roof off the house, signaling the start of another second-story addition.

In this neighborhood of two-bedroom, wartime tract houses it's commonplace for families to add another level of living space atop their original house. Unfortunately, these additions often show little regard for the original shape of the building or the shadows cast across the neighbor's house and yard to the north. Recently, my partner Michael Keenan and I added a second story to a house in Albany that we think is in keeping with the style of the original structure. Our clients, Joe and Denise Lahr, felt that one of the reasons many second-story additions look out of place is that their

height is out of proportion to the original house. To make sure their addition remained in scale with the rest of the building, the Lahrs wanted to keep the highest portion of the roof well back from the sidewalk, with the roofs stepping up as they moved toward the back of the house. The footprint of the original building was basically a rectangle, but it had an appealing roofline because its relatively steep 9-in-12 gable roofs intersected one another in a pleasing asymmetrical geometry of differing ridge heights.

The Lahrs' architect, H. M. Wu, met their needs with a second floor topped with an arrangement of four gables organized around a central hip roof. In plan, the configuration looks a little like a pinwheel (roof plan, facing page). Each gable shelters a distinct part of the addition—one for the sitting room, one for the bathroom, and one each for the two bedrooms. In

the center of the plan, a skylit stairwell leads to the upstairs hallway (floor plan, facing page).

The architect kept the walls as low as possible by giving them a 6-ft. 6-in. plate height. Except for the hallway, each room has a cathedral ceiling, so the low plate doesn't make the rooms feel claustrophobic. The flat ceiling over the hallway is 7 ft. 6 in. high, which leaves enough room above it to run heater ducts for the upstairs rooms.

Hip rafters extending upward from the ridges of each gable meet at the main ridge, forming the tallest of the new roofs. This big hip roof links the four new gables, and repeats the form of the original hipped portion of the roof (photo above).

First, remove the roof—We started our work by stripping the shingles from the south side

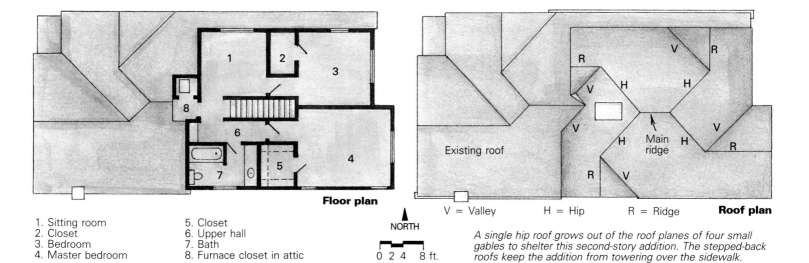

Floor plan

1. Sitting room
2. Closet
3. Bedroom
4. Master bedroom
5. Closet
6. Upper hall
7. Bath
8. Furnace closet in attic

NORTH

0 2 4 8 ft.

V = Valley   H = Hip   R = Ridge   **Roof plan**

A single hip roof grows out of the roof planes of four small gables to shelter this second-story addition. The stepped-back roofs keep the addition from towering over the sidewalk.

of the old roof, letting them fall onto the driveway. Then we pulled out the 1x8 sheathing and the 2x4 rafters. The north side of the roof was very close to the property line, and we didn't want the neighbor's yard cluttered with roof debris. To prevent that, we cut the north side of the old roof into 3-ft. sq. chunks, using a worm-drive circular saw and a 24-tooth carbide blade with wide-set teeth. The wide-set teeth keep the blade from building up a thick layer of asphalt while cutting through the shingles, and the carbide stands up to the many nails that get cut during this kind of work. Since completing this job, we've started using the "negative-rake" carbide blades for this kind of demolition work.

New piers, new beam—The original footings of the house were sturdy enough to take the weight of the addition, but calculations revealed that raising the original wall would shade the neighbor's solar panels. To keep the sun shining on the panels, the Lahrs asked that the addition's north wall be placed 3 ft. south of the existing foundation line. For solid bearing, we needed a beam supported by two 3-ft. sq. by 18-in. deep piers under the house. We used army shovels to excavate the pier holes in the cramped crawl space.

Excavating footings in a crawl space is just plain disagreeable. Fortunately, it's pretty easy to get the concrete under the house with the variety of pumps available today. We used a grout pump with a 2-in. hose for this pour. The small-diameter hose is really maneuverable in a confined space, but it requires small aggregate—⅜-in. pea gravel—and six sacks of cement per yard of concrete (instead of the typical five) to reach a strength of 2,500 psi.

Piers in place, we turned our attention to the original south wall of the house. The second-floor joists of the new addition would bear on the wall's top plate, so we wanted to make sure that it was level. If it wasn't, subsequent construction would require tedious adjustments. Most of the houses around here have settled somewhat, and this one was no exception; we had to remedy the situation.

The existing ceiling joists (over which we were framing the new floor) were rough-sawn 2x4s varying in width from 3½ in. to 4 in. Between each ceiling joist we nailed a pair of 2x4 blocks, face down, using four 16d box nails in each block. The doubled blocks added 3⅛ in. to the height of the top plate. We then notched each ceiling joist to be slightly below the doubled blocks, and ran a continuous plate across all the blocks and joists as a base for the second-story floor joists.

Next we set up the transit on top of the old ceiling joists to check the plate for level. As an aside, I should mention the importance of setting up the transit over intersecting walls so that the tripod legs have solid bearing. If you set up mid-span over flexible joists, you may spend time wondering just when your transit went out of adjustment.

We checked the heights of all the surfaces we were going to build upon, and sure enough, the building was tilted. We were, however, surprised at the severity of the problem. One end of the wall was 2½ in. lower than the other end. We ripped 2x4s into long shims with a bandsaw to level things up. This is the kind of discrepancy that stucco is very good at hiding.

We positioned our new footings in the crawl space so that they fell directly below existing walls of the downstairs bedroom closets. That allowed us to insert the new posts into the walls from the closet sides of the walls, thereby preserving the finished wall surface in the adjoining bedroom and bathroom. The post on the east side of the house tucks into the exterior wall, and we installed it from the outside.

The beam that rides on the three posts is built up from three 2x12s, spiked every 8 in. with 20d nails. Because the beam is wider than the wall that it rests on it, we aligned its inboard edge with the inside of the new wall. The 2x8 joists of the new floor are supported by joist hangers and a ledger nailed to the side of the beam (drawing next page).

Framing the gables—Layout of the stairwell and joists was next, and to nobody's surprise the existing building was not square. After nailing down our plywood subfloor, we squared the new outside walls as best we could by dividing the error as we snapped our chalklines

for the wall. At one end the wall plates project ³⁄₁₆ in. beyond the edge of the subfloor, while they fall ³⁄₁₆ in. inside the edge of the subfloor at the other end. Once again, the stucco would hide the discrepancies.

Before framing the rake walls at the end of each little gable, we snapped chalklines on the subfloor to mark their full scale dimensions, along with all stud and header locations. That allowed us to take direct measurements for the angled studs that intersect the top plates. We positioned the chalkline for the bottom plate to coincide with the plate's baseline so that we could frame the first wall, stand it up and nail it in place, and then use the same layout marks to frame the opposite wall.

In order to keep the sidewalls low, we had to put the window headers on top of the wall plates and affix the rafters to them with joist hangers. The headers are 4x6s with a bevel along one edge to match the 9-in-12 roof pitch (drawing next page). Once the exterior walls were sheathed with ½-in. plywood, we placed our doubled 2x10 ridge beams and began installing the common rafters (bottom left photo, p. 85).

Three pairs of common rafters meet at the main ridge. To set this ridge, we first cut the common rafters that join it, and adjusted their length to accommodate the thickness of the ridge. Then we nailed the tails of the rafters to the wall plates, while aligning the ridge by hand between them. In this manner the height of the ridge was automatically set, and we could then bring a bearing wall from below to carry its load.

As anyone who has done any framing knows, it is especially gratifying to see the walls and rafters in place. All the drudge work in the crawl space starts to pay off when the bones of the building begin to give it shape. On this job, everything had been going well—too well.

After setting the main ridge and some of its common rafters, we noticed that the inboard gable-end wall in the master bedroom had gone out of plumb where it met the exterior wall. When building this rake wall we had been unable to install a post directly under

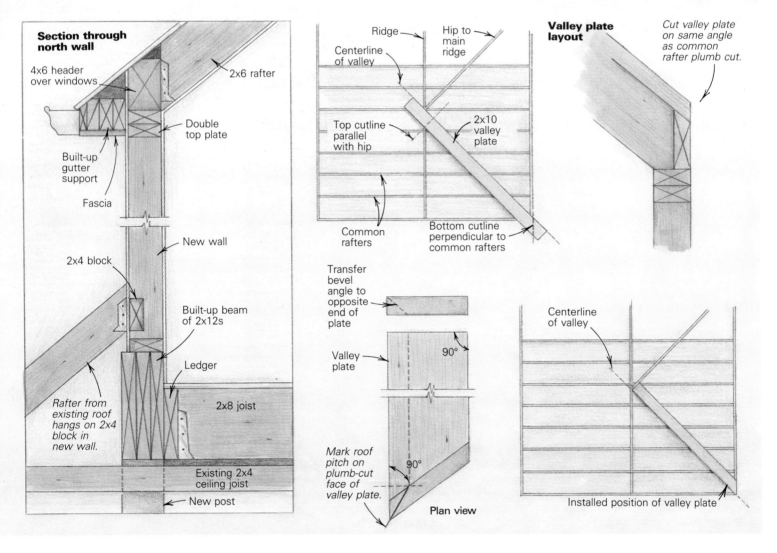

Section through north wall

4x6 header over windows

2x6 rafter

Double top plate

Built-up gutter support

Fascia

New wall

2x4 block

Built-up beam of 2x12s

Ledger

Rafter from existing roof hangs on 2x4 block in new wall.

2x8 joist

Existing 2x4 ceiling joist

New post

Ridge

Centerline of valley

Hip to main ridge

Top cutline parallel with hip

2x10 valley plate

Common rafters

Bottom cutline perpendicular to common rafters

Transfer bevel angle to opposite end of plate

Valley plate

90°

Mark roof pitch on plumb-cut face of valley plate.

90°

Plan view

Valley plate layout

Cut valley plate on same angle as common rafter plumb cut.

Centerline of valley

Installed position of valley plate

its ridge because of plans to install a heating duct there. There were no horizontal top plates to stop the top of the wall from spreading as we added the weight of the rafters in that bedroom, as well as the rafters supporting the main ridge. When we spotted the problem the wall was out of plumb by ¾ in., and we had the uneasy feeling that the error was compounding at a slow but steady pace. We needed a quick and effective fix.

The interior east/west wall of the master bedroom was still plumb, so we braced it with angled studs nailed to the floor and tied a come-along to its top plate. Then we looped the come-along cable around the top plate of the exterior wall. Before winching the wall back toward the house, we placed floor jacks underneath the main ridge of the house and the ridge of the master bedroom. That allowed us to relieve the load on the exterior wall that was driving it outward while we winched the wall back a little past plumb. Then we sheathed the overloaded wall with ½-in. CDX plywood and linked the opposing rafters with 2x4 collar ties a couple of feet down from the ridge. The wall did spring back to about ⅛ in. out of plumb after we removed the jacks and the come-along. That's a discrepancy we can live with.

California framing—With the main ridge set and the frame bolstered against movement, we were ready to install the four hips and four valleys. As shown in the roof plan (previous page), the hips extend from the ridge beam of each gable to the main ridge beam. Because the hips meet their ridge at 45° in plan, we made the double cheek plumb cuts on their ends with the circular saw set at 45° (for more on the mechanics and theory of roof framing, see pp. 58-63).

Our crew uses the "California-style" valley for roofs. Instead of jack rafters from intersecting ridges meeting at a valley rafter, a California valley is built on a 2x plate that lies flat atop the common rafters of one of the gables (bottom right and top photos, facing page). We use this method because it takes less time to build than conventional valleys, and in this case, we wanted as many common rafters as possible in the individual gables because the drywall ceiling is affixed directly to their bottom edges.

To make a California-style valley, we first snapped a line across the tops of the common rafters to mark the centerline of the valley where the two roofs would intersect. Then we laid a 2x10 next to the chalkline, as shown in the drawing above. Using a straightedge, we marked two cuts on the plate. The bottom cut is perpendicular to the common rafters, while the top cut is parallel with the hip rafter. The length is found by measuring in place.

We made the bottom cut with the circular saw set to the plumb cut for a 9-in-12 roof, which works out to 37°. This puts the face of the cut in plane with the sides of the intersecting rafters. On the face of the cut, we marked a 9-in-12 pitch line to represent the plane of the tops of the interesecting rafters. The leading edge of the valley plate should be beveled to this line. The plan view of the plate shows how the pitch line can be transferred to the opposite end of the plate, where you can make a direct degree reading. We typically bevel these plates on our 12-in. bandsaw before nailing them in place to the common rafters. Then we install blocks next to the plate, between each common rafter, as a nailing surface for the roof plywood.

We measured the jack rafters between the valley and the ridge in place. Their tops had the plumb cut angle of the common rafters, but with a 45° cheek cut to meet the hip rafters. Their bottom cuts were level, with the saw set at 37° to match the 9-in-12 pitch of the valley plate.

Before we sheathed the roof we laminated three 2x4s and a 1x4 directly to the walls for a gutter support that doubles as a narrow eave (section drawing above). We beveled the tops of the two outer laminations to be in line with the roof plane and finished the detail with a fascia board affixed to the bottom of the gutter support. □

Alexander Brennen is a partner in Zanderbuilt Construction in Berkeley, California.

Framing details. Before the hip roof was erected, Brennen and his crew framed the four small gables that enclose the upstairs rooms (photo above). The joists in the foreground will carry the ceiling of the hallway. To the left is the skylight well. A California-framed valley relies on a valley plate for jack-rafter bearing rather than on a valley rafter (photo right). The plate is nailed to the common rafters, and its leading edge is aligned with the centerline of the valley. In the photo at the top of the page, the author aligns the lower end of the plate with the valley's centerline, while Michael Keenan snugs the upper end of the plate against a hip rafter leading to the main ridge.

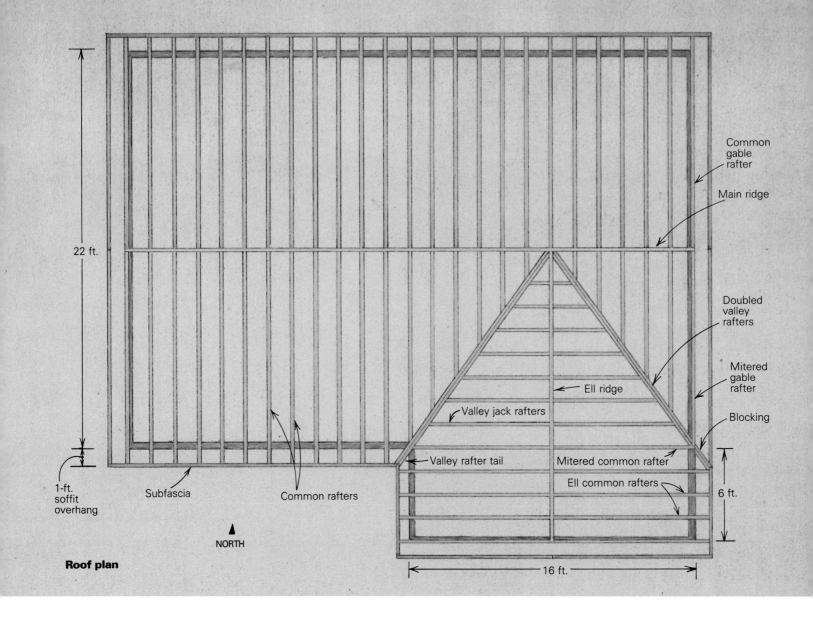

Common
gable
rafter

Main ridge

22 ft.

Doubled
valley
rafters

Mitered
gable
rafter

Ell ridge

Valley jack rafters

Blocking

1-ft.
soffit
overhang

Subfascia

Common rafters

Valley rafter tail

Mitered common rafter

Ell common rafters

6 ft.

NORTH

Roof plan

16 ft.

Valley Framing for Unequally Pitched Roofs

An empirical method that works

by George Nash

The intersection of two roofs with unequal pitches involves geometrical relationships not readily visualized or easily understood. Graphic projections like the method detailed by Scott McBride in "Roof Framing Revisited" (see the article, pp. 9-15) can be intimidating to anyone without substantial framing experience. As for me, I want to frame the roof, not tinker with models, pens, paper and a calculator. In fact, I'm convinced that "fear of ciphering" is so com-

mon that few framers have anything more than a vague notion of somehow using "strings and levels" to lay out complex roofs. This sometimes translates into blundering through, trial-by-error or fudge-and-fix techniques, with hopes that the client doesn't show up until the roof sheathing has hidden the mistakes.

That's how it went for me until a summer when everything I built had a weird roof. I needed a framing method that was fast, accu-

rate, relatively simple and, most of all, non-mathematical. I've forgotten all the trigonometry I never learned in high school, so I'm hopelessly doomed to be a string-and-level man. In the article that follows, I'll describe that method for you and apply it to a house I built that's fairly typical of unequally pitched roof framing.

Purists, or those more mathematically adept than I, may find my methods inelegant, or

Drawings, except where noted: Christopher Clapp

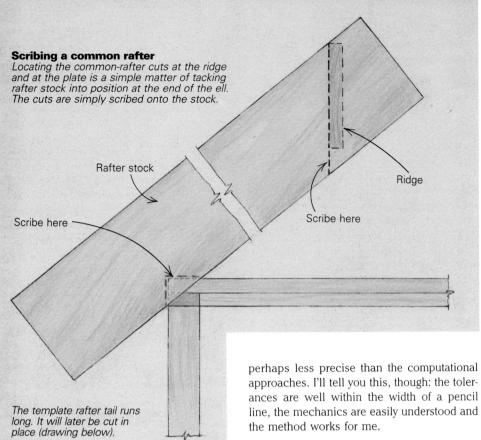

Scribing a common rafter
Locating the common-rafter cuts at the ridge and at the plate is a simple matter of tacking rafter stock into position at the end of the ell. The cuts are simply scribed onto the stock.

Rafter stock

Scribe here

Ridge

Scribe here

The template rafter tail runs long. It will later be cut in place (drawing below).

Rafter-tail layout

Rafter

Level

Framing square

Offcut

Level line

Determining the length and cutting pattern for a common rafter tail on the ell can be done without calculation. Snap a level line on the wall, corresponding to the common rafter level cut on the main roof. Then use a spirit level to transfer this line to the common rafter tail. Slide a framing square along this line until it measures a given distance along the vertical leg; this will be the plumb cut.

perhaps less precise than the computational approaches. I'll tell you this, though: the tolerances are well within the width of a pencil line, the mechanics are easily understood and the method works for me.

First things first—In the project drawings I was given, the L-shape of the Stoecklein house appeared to include a conventional valley, but it didn't. According to the drawings, the ridge was at the same height for both roofs, but the rafter span of the main roof was 22 ft. while the ell span was only 16 ft. The main roof was framed at a 7-in-12 pitch. In order for a smaller span to terminate at the same eave height, the pitch of the ell had to be steeper.

The first rule for framing uncommon rafters is to lay out and install all the rafters that ain't (in other words, do the common rafters first). I'll assume you know how to use a framing square to do this; if not check *FHB* #10, pp. 56-61. After cutting and installing all the rafters on the main roof, I was ready to tackle the ell (drawing facing page).

Installing the ridge—As for framing the roof of the ell, the idea is to work from the top down, which means getting the ridge into place and *then* installing the rafters. The first step was to measure and mark the midpoint of the ell top plate and center a plumbed and braced 2x4 post over it. This would support the outboard end of the ell ridge until the ell's common rafters were installed. The post was cut to the same length as the distance between the top plate and the underside of the main ridge. Working from pipe scaffolding, I transferred the centerline of the ell onto the main ridge to locate the intersection of the ell ridge. (I always use rented pipe scaffolding for roof framing. With two sections and enough staging plank, all but the longest roofs can be framed with minimal movement.)

Figuring the length of the ell ridge was easy. It had to run the full length of the ell, plus half of the full width of the main building, minus one-half the thickness of the main roof's

ridgeboard. In this case, that meant 6 ft. (the ell) plus 11 ft. (half the main roof) minus ¾ in. (half the ridge). So the ell ridge would be 16 ft. 11¼ in. long. After cutting the ell ridge to length and marking out the rafter spacing on it (better now than when it's up in the air), I nailed it into place at the main ridge and atop the ell centerpost. I double-checked to make sure that the end of the ell ridge ended plumb over the gable wall. A temporary diagonal brace run down to the deck held everything in place.

Ell common rafters—Once the ridges and main-roof common rafters were in place, the layout for the ell common rafters was simple: I pinned rafter stock against the end of the ridgeboard and the corner of the wall plate and scribed for the plumb cut and bird's mouth (top drawing, left). No, it's not elegant, but it works perfectly. The position and depth of the bird's mouth followed from the rule that the seat cut should begin at the inside edge of the top plate. The length of the rafter tail will determine how far away from the wall the fascia will be, so the rafter tails on the ell had to be laid out to allow the ell fascia to flow continuously into the main fascia. Rather than including this step in the initial layout, I simply made the plumb cut at the ridge and the bird's-mouth cuts, leaving ample tail stock to be trimmed later. I cut two rafters and tacked them to the ridge to test the fit.

With the two test rafters in place, it was easy to lay out the cuts on their tails. First I leveled across from the bottom edge of a main-roof common rafter tail to the wall itself, as if laying out a horizontal soffit lookout, and then measured the distance from this mark to the top of the wall plate. Returning to the ell, I measured down the wall this same amount and snapped a level line across the wall. Then it was short work with a level and a pencil to extend this line across the bottom of the extending rafter tails; this would be the level cut (bottom drawing, left). To get the plumb cut, I moved a framing square horizontally across this line until it measured a vertical line equal in length to the plumb cut of the main-roof rafters.

It's important to note that if you want the intersecting ridges to be of the same height and the fascias on both parts of the house to line up, the width of the ell soffit will be less than that of the main soffit. If the ell were wider than the main roof, the reverse would be true. If you'd rather have the soffits be equal in width and at the same elevation all around the house, then one of the ridges must be lowered or raised accordingly. Usually these sorts of details are worked out in the design phase. On this job, the difference in soffit width amounted to slightly less than 3 in., which really isn't noticeable.

Although the method I just described will establish the tail cut for either horizontal or pitched soffits, I'd recommend using a horizontal soffit unless the design is beyond your control or changes are not allowed. Horizontal soffit boards and vents are much easier to fit and nail than pitched ones.

About this time I'll usually support the intersecting ridges with a temporary post. Otherwise the ridges could sag as the valleys and their jack rafters are added, and the plumb cuts and lengths of the jack rafters would become increasingly inaccurate. I always check the ridges for straightness, or line them to a string, before laying out the valley rafters.

Finding the valley length—A valley rafter has a lot of cuts and angles to line up, and you'll have a lot of lumber to throw away if one calculation turns out wrong. Fortunately, there's a way to isolate each component and reduce the chances for confusion and error.

Because I always use a subfascia of 2x stock (for the extra support it gives to the soffit), finding the length of the valley rafter isn't tough. First, I nailed the subfascia to all the common rafters around the house. Where the ell intersected the main building, I extended the subfascias to meet at the inside corner (drawings below). Because I had beveled their top edges to match the corresponding roof pitches, it took some fudging with a trim plane to fit the steeper bevel to the shallower one. Some carpenters skip the bevel and simply drop the fascia slightly instead (either method will provide a nailing surface for the edge of the roof sheathing). Then I nailed the intersecting subfascias together. I stretched a string from the outside corner of this intersection to the intersection of the two ridges, right to the top edge; this represented the center line of the valley rafter's top edge (drawings below). Finding the actual rafter length was simple: I just measured along the string.

Figuring the plumb cuts—To find the face angle and the edge angle of the valley-rafter plumb cuts, I used a sliding T-bevel to copy the angle between the string and the ridges. The same angle marked the heel cut of the bird's mouth. It was easy to use a short level and plumb up from the wall plate to the string and then measure the distance to determine not only the heel cut, but the depth to the seat cut of the bird's mouth and the length of the valley from ridge to plate (drawings facing page). The tail cut was simply the same angle repeated where the string crossed the intersecting subfascia. Before making the actual cuts I transferred the angles to a short length of stock and cut a test piece—mistakes on scrap stock are a lot easier to correct.

A doubled valley—A valley rafter on a roof with *regular* pitches calls for a double cheek cut where the valley rafter intersects the ridges and the subfascia. The top edge of the valley rafter then has to be "dropped" just enough to allow roof sheathing to clear it. But the double cheek cut can be complicated, and dropping the valley leaves very little support for fastening the roof sheathing. That goes against the grain of my framing aesthetic—I like plenty of meat to nail into. That's why I double the valley rafter. And if the top edge of each doubled valley rafter is beveled to match the plane of the adjacent roof, the rafter will provide a much better nailing surface for the sheathing (small drawing, facing page).

Of course, two trial pieces with single cheek cuts already made are needed, one for each half of the doubled valley. To find the angle of the top bevels, I lined up each of my trial pieces with the valley rafter center-line string

and held it at the intersection of the two ridges. Then I scribed it where the stock projected above the ridge. This is called backing the valley. Because the resulting angle will be scribed across the face of the compound angle (the ridge plumb cut) and not the square edge of the rafter stock itself, it can't very easily be duplicated with a T-bevel. Instead, I used trial and error—when the cut of the table saw matches the scribe line, I've got the right angle. After the top bevels were cut, I installed the paired rafters and spiked them together.

By the way, the same benefits of doubling the valley rafter apply when it must support a finished ceiling. In that case, a 2x4 ripped to the required width and bevel will furr out the underside of the double rafter for solid nailing, and the finished intersection of the different ceiling planes will be more accurate (small drawing, facing page).

While I left a tail on the doubled southwest valley rafters, letting them intersect the fascia, I dispensed with tails on the southeast valley rafters where the ell shared a common wall with the main-roof gable. Instead, I cut a 45° miter in the plate end of the main roof's gable rafter and did the same thing with the intersecting common rafter of the ell. This way they'd fit against each other at the outside of the wall plate and automatically give the correct height for the center-line string. In lieu of the valley rafter tail, the soffit was fastened to a lookout, and blocking above carried the edge of the roof deck.

By the way, I left the center-line strings in place until all the valley jack rafters were finished. Even a doubled valley rafter will shift with the push and shove of the jack rafters and the weight of the carpenters as they clamber

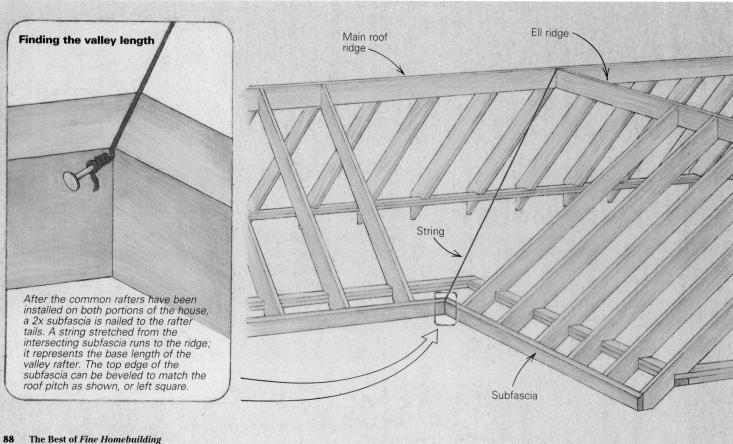

Finding the valley length

After the common rafters have been installed on both portions of the house, a 2x subfascia is nailed to the rafter tails. A string stretched from the intersecting subfascia runs to the ridge; it represents the base length of the valley rafter. The top edge of the subfascia can be beveled to match the roof pitch as shown, or left square.

Main roof ridge

Ell ridge

String

Subfascia

Finding the valley-rafter cuts

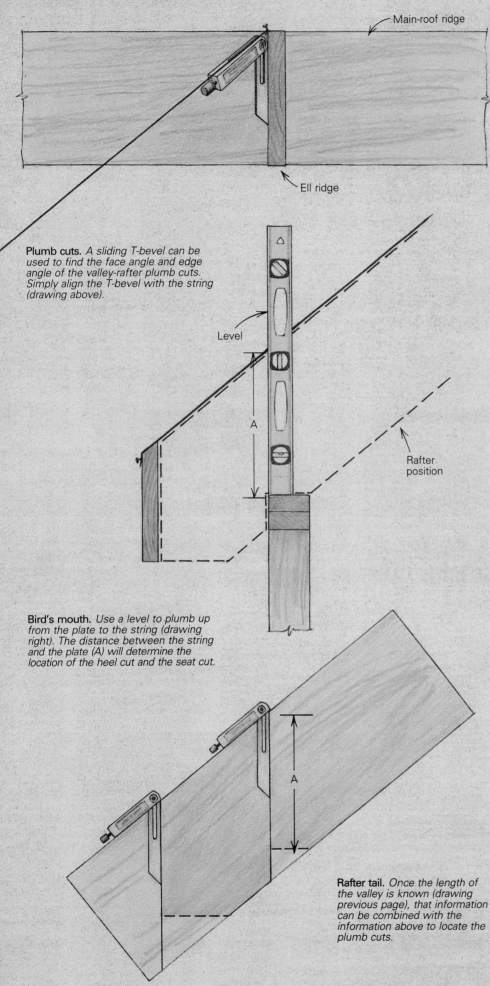

Main-roof ridge

Ell ridge

Plumb cuts. *A sliding T-bevel can be used to find the face angle and edge angle of the valley-rafter plumb cuts. Simply align the T-bevel with the string (drawing above).*

Level

A

Rafter position

Bird's mouth. *Use a level to plumb up from the plate to the string (drawing right). The distance between the string and the plate (A) will determine the location of the heel cut and the seat cut.*

A

Rafter tail. *Once the length of the valley is known (drawing previous page), that information can be combined with the information above to locate the plumb cuts.*

about, especially if the span is long. The string is a convenient guide for constantly checking alignment. Temporary braces may be needed to hold the valley rafter to the line until all the framing is complete.

The valley jack rafters—I have found that the tables of common differences for jack and hip rafters on lines 3 and 4 of the framing square don't always lead to perfect cuts. There are just too many 16ths and smidgens in a real framing job for it to correspond exactly with a theoretical frame. And because I was dealing with an odd pitch on this project, I wanted to derive the common difference (the uniform difference in length between each successive jack) by measuring the actual distance between the first two jack rafters, not by consulting a table. It was string and level time, phase II.

The layout lines for the jack rafters were already marked on both ridges. All I had to do was make a corresponding tick mark on the valley rafter at the right place to find the length and face angle of the jack. I knew that the center of the jack would have to be 16 in. away from the center of the nearest common rafter and be parallel to it, so I was able to use my square to pinpoint its intersection with the valley. (You'll have to eyeball the common rafter for straightness and take out any bows by bracing with a temporary board before you try this.) With the long side of the square resting on top of the common rafter and the short side resting on the valley rafter, I simply slid the square up and down until I located a point on the outside of the valley exactly 16 in. away from the outside of the common rafter (top drawing, next page). This represented the intersection of the valley jack with the valley.

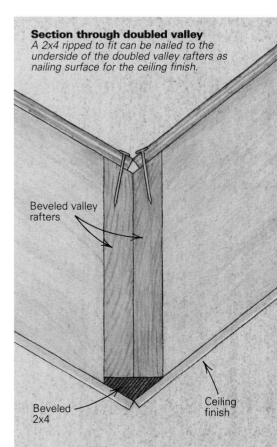

Section through doubled valley
A 2x4 ripped to fit can be nailed to the underside of the doubled valley rafters as nailing surface for the ceiling finish.

Beveled valley rafters

Beveled 2x4

Ceiling finish

Framing Roofs   89

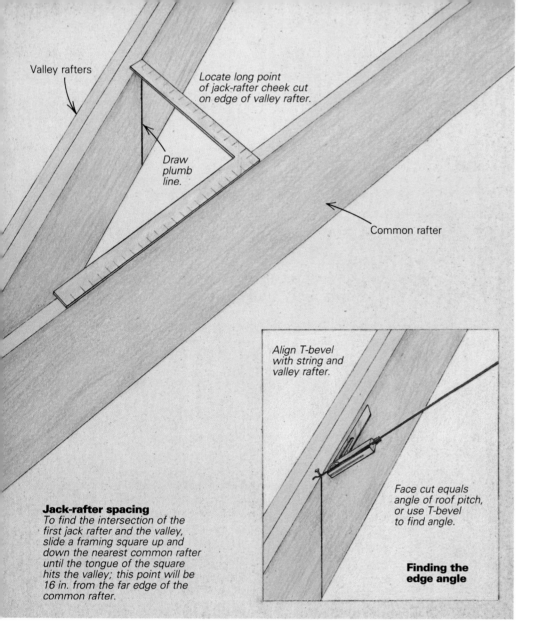

Valley rafters

Locate long point of jack-rafter cheek cut on edge of valley rafter.

Draw plumb line.

Common rafter

Align T-bevel with string and valley rafter.

Face cut equals angle of roof pitch, or use T-bevel to find angle.

Finding the edge angle

Jack-rafter spacing
To find the intersection of the first jack rafter and the valley, slide a framing square up and down the nearest common rafter until the tongue of the square hits the valley; this point will be 16 in. from the far edge of the common rafter.

Then I drew a plumb line down the side of the valley rafter using a short level.

I fastened a string from the top of this mark to the ridge, parallel to the common rafter, and stretched it tightly. Then I aligned the body of the T-bevel with the string and set the blade against the valley rafter; this gave me the angle of the cheek cut. It's important that the blade and the center of the T-bevel handle be in the same plane when lined up to the string; a small twist will give you an incorrect angle. Here again, you'll want to make a trial piece from scrap stock to test the fit before cutting the actual rafter. Because the valley had been doubled up earlier, instead of dropped, the point marked was exact.

To determine the jack-rafter length I simply measured the string from the ridge to the tick mark on the valley rafter. I duplicated the angle of the plumb cut at the ridge on each successive piece using a framing protractor. Making the compound cuts first before laying out the plumb cut at the ridge end ensures a good fit. The string was no longer needed once I found the angles and verified the fit.

After I installed the first jack rafter, it was easy enough to determine the common difference simply by measuring it. All I had to do now was repeat the series of cuts on each remaining jack rafter, reducing the length of each one by the common difference.

To keep the valley rafter in line, the jack rafters are usually nailed home in opposing pairs. But with unequal roof pitches, the cheek cuts are not mirror images and the spacing intervals will not line up across from each other. To avoid throwing subsequent measurements off, I find it easier to work up one side of the valley at a time, marking the position of the next jack rafter by setting the framing square along the edges of the last. I used bracing to keep the jacks from crowding the valley rafter off the center line.

The beauty of this empirical method is that no advance preparation is required before framing can begin. A good rule of thumb is: if you count the jack rafters for one side of the valley as if they were common rafters and add an extra, there won't be much waste. The shorter jacks are usually cut from the leftovers of the longer ones. □

George Nash lives in Burlington, Vt.

Making cheek cuts

The worst thing about jack rafters is making the cheek cut, which is almost always greater than 45°. Sawing through a 2x12 at a compound angle with a handsaw is tedious and tiresome; using a chainsaw is dangerous and usually not very accurate, and I don't have a compound miter saw. Instead, I used applied geometry, some power and a dash of old-fashioned elbow grease.

For example, suppose the edge angle figures out to be 72° and the face angle 35°. I first cut the face angle across the face of the rafter (with the saw set at 90°) and then

tack the rafter to the sawhorse. Complementary angles must add up to 90°, so the complement of 72° is 18°. If I set the saw at that angle and then hold its base against the edge cut itself (perpendicular to the side of the rafter) I can make a 72° cut. Although a 7¼-in. blade will not cut all the way through the angle, what's left is fairly easy to finish with a handsaw. An 8¼-in. or 12-in. circular saw would be handier. I know of no easier way to make these cuts than on a radial-arm saw, which takes more time to set up.

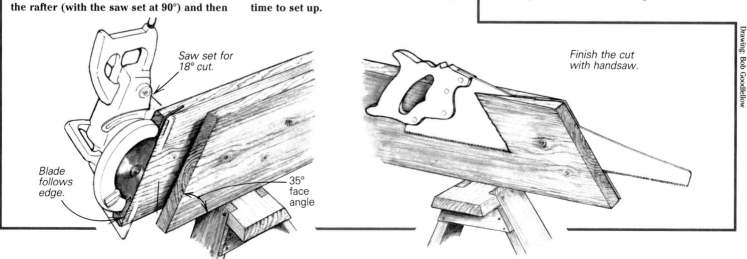

Saw set for 18° cut.

Blade follows edge.

35° face angle

Finish the cut with handsaw.

Drawing: Bob Goodfellow

West Coast Overhang

A framing technique for soffit-covered eaves

by Don Dunkley

Until a few years ago, the typical eave detail on the West Coast was the exposed overhang. When we built ranch-style houses, you could see the rafters and the underside of the roof sheathing (usually ship-lapped boards exposed to view). But the current trend where I build in California's Central Valley is toward the Mediterranean look; more and more of the houses we frame have soffit-covered eaves.

In the past, the crew used to grumble and moan when they got to the soffit-framing part of a job. The framework for our first few attempts consisted of sun-baked 2x4s salvaged from the scrap pile. We scabbed them to the rafter tails and toenailed them to a horizontal nailer affixed to the wall studs. Cobbling together row upon row of outriggers around the eave line took a lot of time, and leveling each one was equally frustrating.

About a year ago, however, I learned of a slick method for installing soffits that uses the fascia to support the outboard edge of the soffit. Our crew quickly put the idea to the test, and the job went so smoothly that we now use the method as standard procedure.

In the groove—For soffit material, we typically use ⅝-in. thick plywood with a resawn face, and slip it into a groove cut in the backside of the fascia (top drawing at right). Plywood this thick is stiff enough to span up to about 30 in. without additional support.

Before cutting any rafters, I calculate the position of the level cut and the plumb cut on the rafter tails. First I step off 1 in. on the fascia (usually a 2x10) to allow for a reveal below the soffit. Next I mark off ¾ in. to allow for the soffit groove. I make the groove ⅛ in. wider than the thickness of the soffit material to give me some wiggle room for inserting the edge of the plywood. The remaining width of the fascia represents the maximum depth of the rafter plumb cut. I always lay out the level cut on the rafter to end up a little bit above the groove in the fascia—½ in. to ¾ in. will do. This makes up for the inevitable variations in framing lumber that can bring the bottom of the level cut into the plane of the soffit. I make all the level cuts on the rafters before stacking the roof.

The top edge of the fascia should be beveled to match the pitch of the roof. Most of the lumberyards that I deal with can provide this service. If yours doesn't, take the time to bevel the fascia yourself. It will give you a consistent line at the roof edge and solid bearing for the roof sheathing.

I use my router, guided by a 1x4 tacked to the back of the fascia, to plow the groove. My Skil model 5000 router is closing in on 20 years old, and it's rated at a mere ⅞ hp. But with a ¾-in. carbide bit it still gets the job done (though it takes two passes to get to the full ½-in. depth of the groove).

I rip my soffit plywood about ¼ in. shy of the overall dimension between the stud wall and the bottom of the groove in the fascia. This makes it easy to slip the plywood into the groove without the free edge hanging up on slight bows along the wall.

To vent the roof we drill trios of 3 in. holes, 3 in. apart. The center holes in each group are 4 ft. o. c. (bottom drawing at right). Our standard practice is to start the center hole in the first trio 10 in. from the beginning of the sheet. Drilled on this layout, the spacing between the holes is consistent from sheet to sheet. To keep the bugs out, we staple insect screen over the holes on the concealed side of the plywood.

The inboard edges of the soffits are affixed to nailers with hand-driven 8d hot-dipped galvanized nails. I locate the bottom edge of the nailer by first leveling across from the bottom of the fascia to the wall. I mark this point, and measure up from it the distance from the bottom of the fascia to the top of the groove. Measuring down from the top plate of the wall to this mark gives me a constant reference point for positioning the nailer. It's a good idea to put building paper on this part of the wall before installing the nailer. (If the wall depends on plywood for shear resistance, it must go on before the nailer.) I let the paper hang down below the nailer far enough so that the siding or stucco crew can tuck their paper under it. If there are openings above the nailer between the stud bays and the soffit, I draft stop them with pieces of plywood.

With this system down pat, my crew can run a substantial amount of soffit in a day's work. The only limiting factor seems to be the time involved in setting up the scaffolding for the predominantly two-story houses that we build. □

Don Dunkley is a framing contractor living in Cool, California.

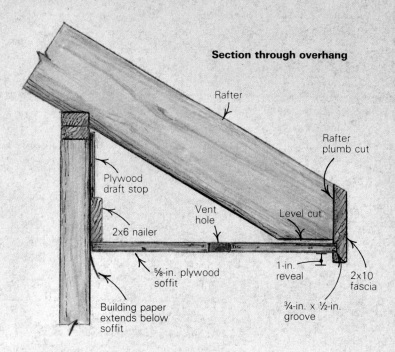

Section through overhang

Rafter

Rafter plumb cut

Plywood draft stop

Vent hole

Level cut

2x6 nailer

⅝-in. plywood soffit

1-in. reveal

2x10 fascia

Building paper extends below soffit

¾-in. x ½-in. groove

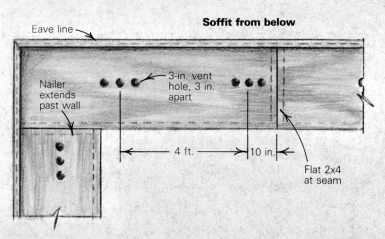

Soffit from below

Eave line

Nailer extends past wall

3-in. vent hole, 3 in. apart

4 ft.

10 in.

Flat 2x4 at seam

Aligning Eaves on Irregularly Pitched Roofs

Making soffits and fascias line up when intersecting roof pitches aren't the same

by Scott McBride

Spicing up a gable complicates framing and trimming. A pair of steeply pitched gables adds curb appeal to this house with a medium-pitch gable roof. Getting the rafter tails to line up required raising the wall plates on the bays, or kickouts.

Pick up a catalog of stock house plans in any supermarket these days, and you'll see that cut-up roofs—roofs with lots of hips and valleys —are back in fashion. The highly competitive new-home market has compelled builders to spice up their roofs with tasty devices such as Dutch hips and wall dormers. The desired effect is curb appeal, which is the elusive but all-important quality that plays to the homebuyer's romantic notion of what a dream house should look like.

As long as all intersecting roof slopes are inclined at the same pitch, framing a cut-up roof can be fairly straightforward: When the slopes are the same, all of the hips and the valleys run at 45° in plan. Consequently, all hip-, valley- and jack-rafter cuts can be made on a simple 45° bevel (the cheek cuts), and only two plumb-cut angles are required: the common-rafter plumb cut and the hip/valley-rafter plumb cut.

However, combining steep-roofed projections with a medium-pitch main roof is a good way to compromise between cost and curb appeal. A roof system usually starts with a main gable, and increasing the gable's pitch dramatically increases material and labor costs. Cosmetic roof features such as dormers are much smaller, so increasing their pitch won't have the same impact on cost as will increasing the pitch of the main roof.

Because the usual purpose of cosmetic features is to lend drama to a home's facade, there may be a strong incentive to make secondary roofs steeper than the main roof, especially on the street side.

While adding steeply pitched features to a medium-pitch main roof might seem like an ideal way to increase curb appeal, it complicates the framing considerably. I recently built a house that has such an unequal-pitch condition, and here I want to talk about some of the difficulties I encountered and how I resolved them.

The particulars of this job—The house is rectangular in plan, except for two rectangular bays, or kickouts, extending 16 in. beyond the front wall (photo facing page). The kickouts are topped with 12-in-12 gable roofs. The main roof of the house has a 7-in-12 pitch. Because of the different roof pitches, the valleys don't run at 45° in plan; they're angled toward the lower-pitch main roof. I had to figure out what that angle was and then how to frame the valley.

The eaves were to overhang 12 in. on both the main roof and the kickout gables, and a sloping soffit was to be nailed directly to the rafter tails. If I built the main wall and the kickout walls the same height, the kickout rafter tails would end 5 in. lower than the rafter tails on the main roof (7-in-12 vs. 12-in-12). That would misalign the fascia boards and the sloping soffits.

Complicating matters further, the rafters for the kickout gables were to be 2x6, while the main-roof rafters were to be 2x8. To get a grip on all of these variables, I headed to the drawing board.

Drawing the cornice section—I always begin roof framing with a full-scale cornice section drawn on a piece of plywood or drywall (drawing right). In this case I drew one cornice section on top of the other—the 7-in-12 main roof cornice section and the 12-in-12 kickout roof cornice section. The superposed drawing provided me with the length of the rafter tails, the location of the bird's mouths, the width of the fascia and the depth of the sloping soffits. From the drawing I also determined how high I'd have to raise the kickout wall so that the kickout fascia would line up with the main fascia.

I began by drawing in the 7-in-12 overhang for the main roof, with the lower edge of the 2x8 rafter starting at the inside corner of the wall plate, the typical location for a bird's mouth. Underneath the 2x8 tail I drew the main-roof soffit. From the point where the face of the soffit meets the back of the fascia, I drew a line at a

12-in-12 pitch to represent the more steeply pitched kickout soffit. Next, I drew a parallel 12-in-12 line from the top end of the 2x8 tail of the main roof; this line represented the top edge of the kickout rafter tail. I now had the top and the bottom edges of my rafter tails aligned at a point 11¼ in. away from the wall. (The ¾-in. thickness of the fascia would increase the overhang to 12 in.)

Next, I drew the 5½-in. width of the 2x6 kickout rafter. The remaining distance between the

Additional top plates provide bearing for the kickout's 2x6 rafters while picking them up enough to line up the fascia boards.

Superposing main-roof and kickout-roof cornices

This drawing shows how the wall plate was built up to catch the kickout rafters, which made eaves alignment possible. Starting from the top inside edge of the main wall plate, the author drew the main rafter at a 7-in-12 pitch, then added the 2x6 kickout rafter at 12-in-12 in line with the top end of the main-rafter tail. The bottom of the 2x6 rafters required shims so that the soffits could line up.

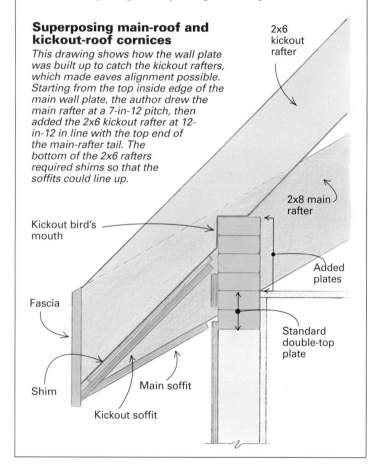

2x6 kickout rafter

2x8 main rafter

Kickout bird's mouth

Added plates

Fascia

Standard double-top plate

Shim

Main soffit

Kickout soffit

lower edge of the 2x6 and the back of the kickout soffit is made up by shimming. The shims cover the underside of the 2x6 kickout tails and extend up along the lower edge of the kickout barge rafters to keep the eave soffit flush with the rake soffit.

At this point I could see roughly how much I needed to raise the kickout wall plate. The exact elevation of the kickout bird's mouth wasn't critical because it's in the attic above the second-story ceiling. I built up layers of 2x4 until the raised plate gave good bearing for the kickout rafters (photo and drawing, left). With the kickout rafters sitting higher on the wall than the main-roof rafters, the fascia and the sloping soffit could flow in a smooth line from one roof to the other.

Simplifying valley construction—There are two methods of building valleys. The first, called a framed valley, employs a valley rafter that supports jack rafters coming down from both intersecting roof surfaces. A simpler approach, known as a California roof or a farmer's valley, is to build the main roof all the way across and frame the intersecting roof on top of it. Instead of valley rafters, you nail valley boards flat on the main roof and frame jack rafters for the smaller roof only (photo p. 94). (For a description of the California approach, see pp. 78-81.)

Because a California roof typically sits on the main-roof sheathing, it can't be used if the smaller roof will have a cathedral ceiling. But there were no cathedral ceilings in this house, so I could use the California approach to frame this roof.

Building a California roof simplified the framing significantly because an unequal-pitch valley has several peculiar traits. First, given that the overhang is the same for both roofs, the valley will not cross over the inside corner where the walls intersect, as is usually the case. Rather, it will veer toward the roof with the lower pitch. A valley rafter's location would have to be figured out beforehand from studying a plan view of the roof framing.

Furthermore, an unequal-pitch hip or valley rafter requires two different edge bevels at the point where it hangs on intersecting ridges or headers. One of the bevels will be sharper than 45°, so it cannot be cut with a standard circular saw. With the California roof, I avoided the problem of dissimilar edge bevels and the hassle of locating valley rafters.

I located the off-center valley simply by snapping lines on the main-roof sheathing. First, I in-

One roof framed on another. A California roof features valley boards that lie flat on the main roof sheathing. As opposed to a valley rafter, there's only one set of jack rafters to cut and one cheek-cut setting on the circular saw.

stalled the kickout ridge and its common rafters. Then, at the peak of the kickout gable, I anchored a chalkline, stretched it diagonally across the kickout common rafters and marked a point somewhere near the eaves where the chalkline hit the main-roof sheathing. The top of the valley occurs where the kickout ridge dies into the main roof, so, by striking a line across these two points, I located the valley boards. I beveled the edges of the valley boards and nailed them flat on the main-roof sheathing.

Mitering the soffits—Once I had the kickout jack rafters and sheathing in place, I turned my attention to the cornice. Thanks to the raised plate on the kickout wall, all of the rafter tails lined up, and the fascia flowed smoothly around the corner. But now I had to make the sloping soffits do the same thing.

The main question was at what angle should the soffit boards be cut to create a clean miter at the inside corner (where the main wall and the kickout wall intersect). I could cut some scrap pieces with 45° face cuts and continue adjusting the angle by trial and error until I had a good fit. Or I could stay on the ground, figure out the angles on paper and install the soffits on the first try.

I opted for method two, and I accomplished this through graphic development. Graphic development is a way of taking a triangle that occurs in space, such as the gable end of a roof, and pushing it down on a flat surface where it can be measured accurately. All you need is a pencil, some paper, a framing square and a compass. A stubborn disposition helps, too.

To figure out the miter angles of the soffits, I began by drawing a plan view of the wall lines and the fascia lines (drawing facing page). I then drew in plan views of the main-roof rafter tail and the kickout rafter tail, each perpendicular to its respective wall. In addition, I drew elevation views of the same rafter tails. I used the numbers 7 and 12 on the framing square to draw the main-rafter elevation view and 12 and 12 for the kickout elevation view. As the plan view and the elevation view of each tail cross the wall lines, they show the vertical rise of the tails: 7 in. for the main roof, 12 in. for the kickout.

Next, I needed to draw the valley. Remember, it doesn't run at 45°. I already had the end of the valley: the point where the fascias intersected. What I needed was another point farther up the valley so that I could draw the valley line. Because I already had drawn the 12-in. kickout elevation view, I decided to find the point where the valley rises 12 in.

First, I extended the kickout-wall line in the direction of the valley. At this line the kickout roof rises 12 in. above the fascia, so the line represents plan views of both the kickout wall and the kickout roof's 12-in. rise line. Then, I extended the main rafter tail and drew a perpendicular line showing the plan view of the main roof where it rises 12 in. above the fascia. The intersection of both of the 12-in. plan-view rise lines is a point on the valley, and, by connecting this point with the inside corner of the fascias, I drew the valley in plan.

To determine the miter angles (or face cuts) for the soffit material, I used a compass to swing the 12-in. high elevation view of each common rafter tail directly over the plan view. This point represents the actual length of the rafter where it rises 12 in. (as opposed to the foreshortened length of the rafter when seen in plan).

Then, I drew lines perpendicular to the plan views of the rafters at the points where my compass intersected them. These are labeled elevation lines on the drawing. Next, I intersected the plan line of one soffit with the elevation line of the other soffit. I connected these intersecting points to the inside corner of the fascia, giving me the angle of each soffit's face cut.

Try to imagine the inside soffit edges rising up while the outside soffit edges remain "hinged" along the fascia line. When the inside edges have risen 12 in., the soffits touch along their face cuts, forming a valley, or as viewed from below, an upside-down hip (photo facing page). The meeting takes place directly over the plan view of the valley line.

The face cut for the kickout soffit was the same as the face cut for the kickout roof sheathing, which happened to be 1x6 but could just as easily have been 4x8 sheets. The sheathing sits on top of the rafters, and the soffit hangs below. Otherwise, they're the same.

Because of its steeper pitch, the kickout soffit dies square into the house for a short distance before mitering with the main-roof soffit. I could tell from the graphic development where to cut the kickout soffit along the main-wall line. The pieces fit on the first try.

Joining mitered soffits—The edge bevel for the soffits wasn't critical because the backside doesn't show. I just cut them at 45°, which undercut the pieces more than necessary and assured a tight miter. Fastening the intersecting soffits presented a problem, however, because I didn't have a valley tail to nail the ends of the soffit plywood to. (That would have been the only good reason for using a framed valley here instead of a California valley.)

To solve the problem, I connected the main-roof soffit and the kickout soffit along their intersection with a beveled 2x4 backerboard. I beveled the 2x4 to match the angle of the valley trough. The 2x4 backerboard didn't have to fit tightly between the fascia boards and the house because the ⅜-in. soffit material was pretty stiff. Instead, I just cut the ends of the backerboard for a loose fit and pulled it against the soffits with galvanized screws. □

Scott McBride, a contributing editor at Fine Homebuilding, *lives in Sperryville, Va., and is a Class-A building contractor. Photos by the author.*

Figuring out the soffit angles

Sloping soffits make for tricky miters. One problem with this roof was getting the sloping soffits of two different roof pitches to flow smoothly around the inside corner. The miter mirrors the offset valley, so the kickout soffit dies square into the main wall. A beveled 2x4 provides backing along the miter joint.

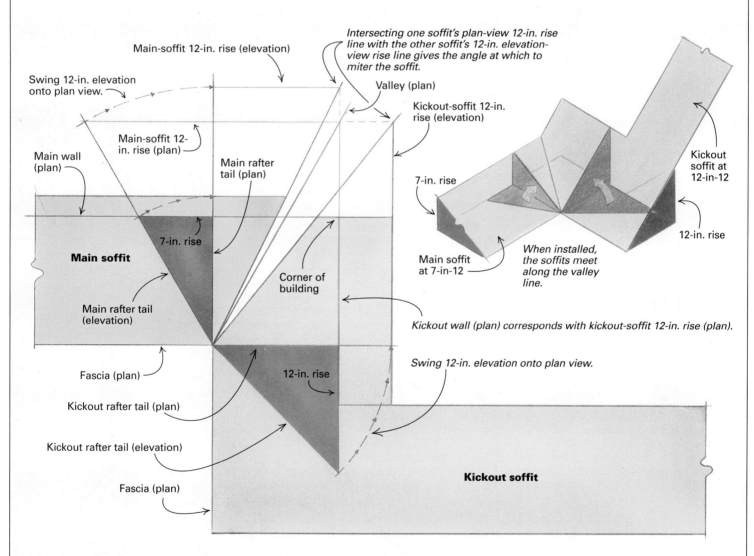

Main-soffit 12-in. rise (elevation)

Intersecting one soffit's plan-view 12-in. rise line with the other soffit's 12-in. elevation-view rise line gives the angle at which to miter the soffit.

Swing 12-in. elevation onto plan view.

Valley (plan)

Kickout-soffit 12-in. rise (elevation)

Main-soffit 12-in. rise (plan)

Main wall (plan)

Main rafter tail (plan)

Kickout soffit at 12-in-12

7-in. rise

Main soffit

12-in. rise

Main rafter tail (elevation)

Corner of building

Main soffit at 7-in-12

When installed, the soffits meet along the valley line.

Fascia (plan)

Kickout wall (plan) corresponds with kickout-soffit 12-in. rise (plan).

Kickout rafter tail (plan)

12-in. rise

Swing 12-in. elevation onto plan view.

Kickout rafter tail (elevation)

Kickout soffit

Fascia (plan)

Combining plan views and elevations.

The challenge was to miter the soffits so that when they're nailed to the rafters, the soffits intersect along the valley line. The solution was found through this graphic development (lower drawing), which shows the soffits lying flat on the paper. The first step was to draw the walls and the fascia. Next came plan views and elevations of the rafter tails. Then, with a compass, these elevations were swung back onto their plan views, which gave the widths of the soffits. Finally, the angle of the soffit miters came from intersecting the plan view of one soffit at a given rise—in this case, 12 in.—with the 12-in. elevation line of the other soffit. When the soffits are pitched at 12-in-12 and 7-in-12, they come together directly above the valley line in plan (upper drawing).

Falling Eaves

One carpenter tries five methods of framing a tricky roof

When an angled bay is capped by an extension of the main roof, the rafters that sit on the angled walls require some careful cuts.

by Scott McBride

Roof framing is tricky enough when the walls are plumb, level and square, but when the rafters intersect an eaves wall that's angled, well, it'll drive a person to thinking. The first time I encountered this condition was with a 45° angle bay extending from a single-story exterior wall (photo above). Instead of having its own separate hip roof, broken up into the usual three planes, the roof of this bay was simply to be an extension of the main roof plane, intersecting the diagonal walls of the bay on an angle in both plan and elevation.

The cornice in this situation runs at some oblique angle in plan (usually 45°) while it falls in elevation. I call this condition "falling eaves," as opposed to regular eaves, which run level.

I improvised my way through that job, thankful that only a few rafters were affected. But when the time came on another house to frame a gable roof with all four of its corners lopped off in this fashion, I decided to study the problem carefully. I ended up with five diferent solutions.

An octagonal room—I was framing a new house that featured an octagonal room extending above the main roof (photo left, facing page). Unlike most octagonal roofs, which have eight roof planes coming to a point at the peak, this roof was essentially a gable, with only two roof planes meeting along a ridge. Falling eaves were located where the angled walls of the octagon intersected this roof.

I began by framing the two gable walls and the two regular eaves walls. To illustrate my method for calculating the heights of these walls, I'll simplify the dimensions a bit. Let's say the level eaves walls were 8 ft. high, and the run of the roof was also 8 ft. (half of a 16-ft. span). Angled walls chop off the four corners of the room (drawing, p. 98), extending in 4 ft. from what would have been a square corner. In that 4 ft. of horizontal run, the 9-in-12 roof rises 4 in. x 9 in., or 36 in. total. The height of the gable wall at its outside corners (the lowest points) would therefore measure 8 ft. plus 36 in., or 11 ft. total. Actually, there's an adjustment that I had to make here, which I'll discuss momentarily.

From its outside corners, the gable wall rises 36 in. over the 4 ft. of run. That would make the height of the gable wall at the peak 11 ft. plus 36 in., or 14 ft. total. But there was a further complication in calculating the height of the gable walls. The calculations just given start at the outside corner of the eaves plate. This point lies on the measuring line, which runs somewhere down the middle of the common rafter. But the gable walls needed to support lookouts for a framed rake overhang (photo top right, facing page). The lower edges of these lookouts line up with the lower edges of the common rafters, in a plane below the measuring line. Consequently, the gable-wall top plates had to be lowered by an amount equal to the vertical depth (heel cut) of the common-rafter bird's mouth.

Falling headers—After tipping up the level eaves walls and the gable walls, I connected them with four sloping headers, which I'll call falling headers. Later, I added 2x6 studs under these headers to frame the angled walls. As with the angled walls of the bay described earlier, the top edges of these falling headers are similar to the top edge of a valley rafter. Because the header travels a horizontal distance of 4 ft. perpendicular to the level eaves wall, I knew that its diagonal run in plan would be 17 in. multiplied by four. The hypotenuse on a right triangle with 12-in. sides is 17 in. (actually it's 16.97). The rise in each of those 17-in. diagonal increments of run had to be the same as for each 12 in. of common-rafter run (namely, 9 in.) in order to keep the header aligned with the roof. To find the actual length of the falling header on its long face, therefore, I stepped off 9-in-17 with the square four times. (A shortcut would have been to multiply by four the number listed under 9 in the length-of-hip/valley rafter table on the framing square.)

Finding my calculated length precisely consistent with the field-measured distance from level eaves wall to gable wall (well, close enough), I drew parallel plumb cuts on the header's outside face at both ends. Through these lines I made opposing cheek cuts, with the blade of the circular saw tilted at 45°. I made four of these headers and spiked two of them into their adjoining walls with 16d nails (I had different plans for the other two headers).

Lopping off the corners. This octagonal room is sheltered by a gable roof that has its four corners lopped off. To find the best way to frame the intersection of the rafters and the sloping, angled eaves walls, the author framed each of the four corners a different way. Photo by Kevin Ireton.

Gable overhang. The gable walls were shortened an extra few inches to allow for 2x10 lookouts, which cantilever beyond the walls to form the overhang. This framing assembly is called a ladder.

It's not supposed to line up. At its lower end, the header in the photo below aligns with the eaves wall. Because the gable wall was lowered to allow for the lookouts, the header protrudes above the gable-wall plate to run parallel with the roof.

Method #1: raised block. Nailing triangular blocks on the sloping, angled headers created a level surface to seat the rafters on. But this required a deep heel cut on the bird's mouth and weakened the rafter overhang.

Because the outside corners of the falling headers were aligned with the level eaves plates at their lower ends, they could not align with the gable-wall top plates at their upper ends and remain parallel to the roof surface. This is because the gable-wall top plates were recessed below the roof surface by the full vertical depth of the rafters (to make room for the lookouts). However, the eaves-wall plates were recessed below the roof surface by the raising distance (the vertical depth of the rafter above the plate). The raising distance takes up only part of the vertical depth of the rafter, with the heel cut of the bird's mouth taking up the remainder (drawing detail next page). Therefore, the top ends of the falling

headers protrude above the gable-wall top plates (middle photo, right). The height of the protrusion is a vertical distance equal to the heel cut of the common-rafter bird's mouth.

I had several ideas about how I might frame rafters into these falling headers. To find out which was best, I decided to frame each of the four corners of the room in a different way.

The raised-block method—Because I'm used to seating the lower ends of rafters onto a level surface, I first tried building up level bearing on one of the falling headers by adding triangular blocks (bottom photo, right). To lay out the spacing for the jack rafters, I pulled 16-in. centers from

Method #2: the notch. Another way to create level bearing for the rafters was to cut a notch in the header with a handsaw. Although it worked, this method did not provide much bearing surface for the rafters.

Method #3: beveled header/level seat cut. Here the falling header is beveled, as is the seat cut of the rafter's bird's mouth. The intersection between the rafter and the header is a level cut (above left). The bevel angle of the header was determined by the top of the eaves-wall plate where the two intersect (above right). The author scribed a scrap block in place to determine the angle. This method provided good bearing and was the easiest to nail.

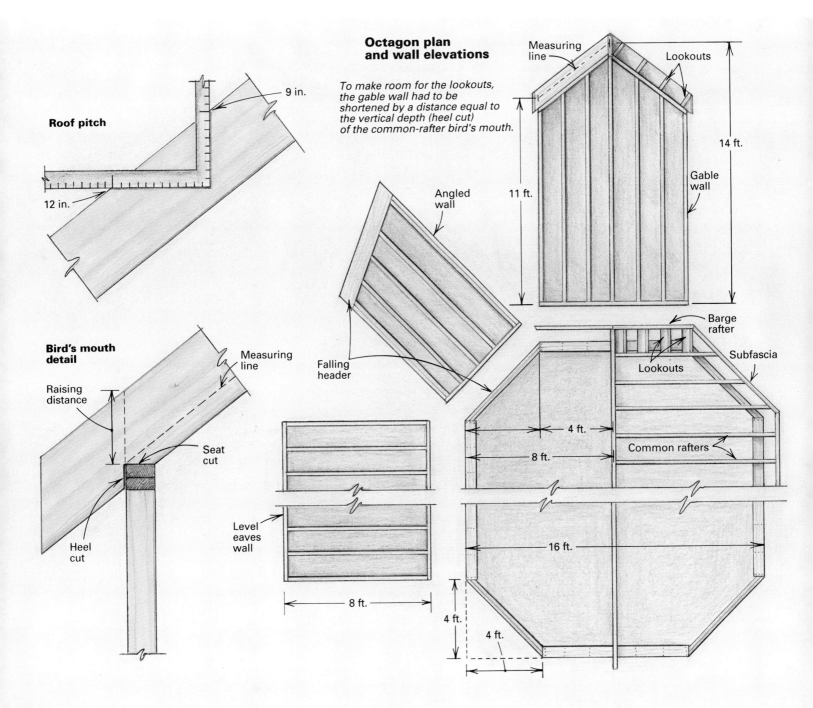

Roof pitch

9 in.

12 in.

Octagon plan and wall elevations

Measuring line

Lookouts

To make room for the lookouts, the gable wall had to be shortened by a distance equal to the vertical depth (heel cut) of the common-rafter bird's mouth.

14 ft.

11 ft.

Gable wall

Angled wall

Bird's mouth detail

Measuring line

Raising distance

Seat cut

Heel cut

Falling header

Barge rafter

Subfascia

Lookouts

4 ft.

8 ft.

Common rafters

Level eaves wall

16 ft.

8 ft.

4 ft.

4 ft.

Method #4: beveled header/no bird's mouth. Here the header is beveled in plane with the bottom of the rafters. There is no need for bird's mouths; the rafters simply bear on the headers (above left) and are held in place by toenails. The bevel angle for this header was determined by the top of the gable-wall plate where the two intersect (above right). The bottom end of the header is recessed below the level eaves-wall plate.

Method #5? From this view, you can't tell whether the header is notched for the rafter or vice versa. The author tried the former but thinks the latter method, which he tested on a model, might be the best.

the nearest common rafter, first making sure the common was straight. The blocks themselves were laid out using the numbers 9 and 17 on the rafter square. With the top of the block presenting a level surface, I could put an ordinary 9-in-12 square seat cut on the jack rafter (cutting on 12). The heel cut of the bird's mouth for these rafters, as well as for the rafters on the other three corners, was made by laying out the standard 9-in-12 plumb cut (cut on 9) on the face of the rafter but cutting it with the circular saw tilted 45°.

As you can see in the photo, the bird's mouths for the blocked-up rafters had to be cut quite deep. This was necessary because the triangular blocks protruded above the top edge of the falling header, which reduced the raising distance and increased the vertical depth of the bird's mouth. Structurally, this weakened the overhang.

The notch method—Another way of producing a level surface on which to seat the rafters was by notching the header. First I reasoned that any line drawn square across the edge of the un-backed header would be a level line. After squaring such a line across the header's edge, I extended a level line on the inside face of the header from the point where the squared-across line hit the header's inside face (photo left, facing page). Along this line, I measured off the 45° thickness of a double 2x—about 4¼ in. From that terminus, I plumbed up to the top inside edge of the header. From where the plumb line struck the corner, I connected back to where I started from on the outside face of the header. These three lines described the two handsaw cuts—one plumb, the other level—that I needed to make the notch.

The bird's mouth to fit these notches was essentially the same one used for the raised-block method, except that it didn't need to be cut extra deep. Because the outside corner of the notch lines up with the outside corner of the level eaves plate, I used the standard raising distance.

Although leaving the strength of the rafter un-compromised, the notch method offers a small bearing surface, which could be a problem with long or heavily loaded rafters.

Beveled header/level seat cut—On one side of the octagon, I used a header backed (meaning that its top edge was beveled) so that its intersection with the seat cut of the rafter was a level line (middle photo, facing page). To find the correct backing bevel, I took a 2x scrap and put a 9-in-17 plumb cut on it sawn at a 45° blade tilt. The scrap mimicked the cheek cut on the end of the header. Holding the cheek cut of the scrap vertically against the end of the level eaves-wall top plate, I scribed a level line across the plate onto the end grain of the scrap. This showed me how much to take off the inside edge of the header (right photo, facing page). Using a worm-drive saw equipped with a rip fence, I beveled one of the headers that I hadn't already installed and then spiked it in place.

To fit the rafters to this header, I laid out the standard common-rafter bird's mouth and made the heel cut (plumb cut) of the bird's mouth with the saw blade tilted at 45°. The seat cut was also made with the blade tilted, but not at 45°. For this I had to use the plumb-cut angle of the common rafter. This method provided good bearing and was the easiest to nail. The backing operation, however, took some time.

Beveled header/no bird's mouth—To round out my experiment, I backed the last header so that its top edge would lie parallel to the roof plane. The big advantage of this method was that the rafters required no bird's mouth whatsoever, so there was no need to calculate precise rafter length (I cut the rafter tails in place later). I was able just to lay the rafter stock down on the mark and toenail (photo above left). I once saw ordinary rafters framed the same way, sitting on walls with tilted top plates, but I suspect that this con-

nection might slip over time (if used with ordinary stud walls) because of the thrust of the roof. In this case, however, I felt that plenty of spikes driven into a beefy header (doubled 2x10s) would adequately resist the lateral load.

To determine the header's bevel, I used the same scrap-block trick I had used for the preceding method, except that I scribed it against the gable-wall top plate rather than against the level eaves-wall top plate (middle photo, above). Unlike the other three falling headers, which are aligned with the level eaves plate and protrude above the gable-wall plate, this header aligns with the gable-wall plate at the top and is recessed below the level plate at the bottom. It's offset from the roof surface by the rafter's full depth rather than by just the raising distance.

And the winnner is—You may be wondering which method I like the best. Well, the beveled-header/no-bird's-mouth method was the easiest. But given my concerns about the rafter-to-plate connection slipping over time, the beveled-header/level-seat-cut method is probably the best of the four I tried.

Since completing the project, however, I have thought of another method. As I looked at a photo of rafters installed using the header-notch method (photo above right), I realized it's impossible to tell from the uphill side whether the rafters are let into the header or vice versa. Instead of notching the header, I could have made a sloping beveled seat cut on the jack rafter's bird's mouth that would mate directly with the edge of the unbacked header. I developed the angles for this method on paper and tested them on a scale model. It works.

Then there was that guy from California who told me he just uses metal framing anchors... □

Scott McBride is a contributing editor of Fine Homebuilding *and lives in Sperryville, Va. Photos by author except where noted.*

Framing a Cold Roof

Preventing roof leaks caused by ice damming

by Steve Kearns

When I moved to Idaho's snow country after living in southern California and Hawaii, I had to learn a number of regional differences in construction. For one thing, it gets colder here, so we use 2x6 studs (at least) to accommodate R-19 insulation in the walls; R-30 is the minimum in our ceilings. I also learned to build what's known as a cold roof. Such a roof, paired with a Boston ridge vent (more on this later), prevents ice damming and roof leaks.

The scenario for a leaky conventional roof in snow country goes like this. With temperatures below freezing, snow accumulates on the roof over a vaulted ceiling. Even if the roof is well insulated, enough heat escapes to melt the snow, causing water to run down the shakes, and when this water hits the unheated portion of the roof (the cold eaves), it freezes. This process happens over and over until ice dams and picturesque icicles form as water drips from the eaves. The icicles may be pretty, but big ones are evidence of a problem roof. When ice dams get big enough, they can cause the melted water to work its way back under the shakes and leak into the house. A constant freeze/thaw cycle resulting from uneven day and night temperatures can also contribute to this ice build-up.

Builders deal with the problem in various ways. Some run a waterproof sheet over the lower portion of the roof, under the roofing. W. R. Grace & Co. (Construction Products Division, 62 Whittemore Ave., Cambridge, Mass. 02140; 617-876-1400) makes a sheet called "Ice & Water Shield" that can be applied directly to the roof decking.

You might think that metal roofing would seal out leaks caused by ice damming, but it doesn't. Mike Kimball, owner of Sun Valley Roofing, reports that metal roofing used on 4-in-12, 5-in-12, and 6-in-12 standard roofs can, and does, leak when ice dams form. On steeper pitches, the slick metal serves as a water barrier only because it encourages built-up snow to slide off (to crush unsuspecting bystanders, flower beds or car hoods...but that's another story). A product like Ice & Water Shield only treats the symptom, though; it doesn't cure the ice-dam problem. To do that we build a cold roof.

Framing the cold roof—The principle of the cold roof is simple: build a double-layer roof

that "breathes." A continuous air flow between the layers from eave to ridge takes any heat escaping from the house and exhausts it out the ridge. This keeps the outermost roof layer "cold" over the whole roof and prevents the ice damming that occurs on a conventional roof. We use 2x4 sleepers to separate the roof layers.

Construction of a cold roof is fairly straightforward. After the rafters of the lower roof have been sheathed, we snap two parallel chalklines about 3 in. apart on the plywood, somewhere over the eave overhang, and cut this strip out with a circular saw. This provides an avenue for air to travel up from the soffit area and into our cold roof. Next we put down 30-lb. felt over the whole roof, cutting it out at the soffit-vent strip. This felt layer protects against any condensation that may occur between the roof layers; it is not intended as a rain or snow barrier. Any condensation will leak out the soffit vent, and hasn't been a problem.

After the felt is down, we stack 2x4 sleepers on edge directly over the rafter layout and run them up the entire length of the roof. This will provide a 3½-in. air space up the roof. I haven't been able to find an engineer who has calculated exactly how much air volume is needed

to keep a cold roof cold, but experience tells me that a 2x4 put down flat doesn't work well enough, while one on edge does. In the course of writing this article, I spoke with Professor Ronald Sack, director of the University of Oklahoma School of Civil Engineering, and he confirmed that analytical predictions for the performance of a cold roof are nonexistent. Ongoing research may, however, yield more precise answers soon. For more technical information on this topic, he suggested "Approximate Analysis of a Double Roof," in the journal *Cold Regions Science and Technology*, volume 16, 1989.

What engineers *can* give us (and some building inspectors require), is a nailing and fastener schedule for attaching cold-roof sleepers to the main roof. I've heard horror stories of a cold roof sliding completely off a house in Mammoth Lakes, California, and our inspectors would rather that didn't happen here. Mike Bouiss, of Bouiss & Associates, P. E., developed the nailing schedule we use. We nail A-35 clips (Simpson Strong-Tie, P. O. Box 1568, San Leandro, Calif. 94577; 415-562-7775) to the plywood directly over the rafters and then put each 2x4 sleeper over the clips, covering the side of the clip that lies flat on the sheeting. Then we nail through the vertical side of the clip into the side of the sleeper. A small aside—U. S. Nail Company (1840 National Ave., Hayward, Calif. 94545; 415-785-7443) makes a heat-treated joist-hanger nail for the Hitachi nailgun that speeds this work (and other metal-framing connector work) immeasurably. After sleepers have been laid over the whole roof, they are skip-sheathed and shaked just like for a conventional roof—up to the Boston ridge vent, that is.

One variation of the cold roof involves an alternative to cutting a vent strip in the plywood sheathing over the soffit. In this case, we extend the sleepers 3 in. or more past the fascia line of the lower roof (bottom photo facing page). The ends of the sleepers can then be finished off with a separate fascia. Whatever method you choose, however, it's important to screen the vent to keep out birds and bugs. We buy our screening in rolls 6 in. wide and staple it in place. A light-gauge metal soffit vent is available in 8-ft. lengths (with a galvanized or bronze finish); it's easy to apply, but it's only 2 in. wide and the perforations cover only about 50% of the vent area. Figuring that these vents might overly restrict air flow, we stick to screening.

The Boston ridge—The Boston is essentially a raised ridge vent with an overhang that prevents the vent from being clogged with snow. It can be applied either before or after the shakes are on, but we prefer to do it before. We attach it to the skip sheathing because that provides a more secure connection.

There are a number of designs for Bostons—some people take 2x8s, plumb cut them where they meet at the ridge, and put a 3½-in. by 5-in. notch in the lower end to provide the protected vent (bottom drawing next page). Another way (which we find faster) is to build 2x4 ladders along the length of the ridge; we stack the two wider ladders on top of the two shorter ones to provide a protective overhang for the vent (top drawing next page). The cross members are 4 ft. o. c., and the upper ladders are simply spiked into the lower ones. Don't forget to staple screening over the entire vent opening—to birds and bees, these Boston ridges look like dream homes.

The gable ends and the eaves of a Boston ridge are normally finished with fascia to complement the regular roof. We sometimes double or triple the fascia along the edges of the regular roof, which gives our roofs a massive look.

One design issue is whether to bring the Boston ridge all the way out to the end of the gable's eaves or to stop it at the house line. The roof doesn't need to be vented over the already cold eaves, of course, but some architects prefer the look of a full-length Boston,

The sleepers on adjoining roof planes at hips and valleys should be oriented to channel air upward. Nothing should obstruct the flow of air from soffit to ridge.

In this application of a cold roof, the sleepers extend several inches past the fascia of the underlying roof to form a protected vent area. Screening (just visible in the foreground of the photo above) has been stapled in place to keep bugs and birds from invading the roof later on.

The Boston ridge (photo above) used for a cold roof is a site-built variation of the standard ridge vent, but with an overhang that prevents it from clogging with snow. Two methods for framing the vent are shown in the drawings below.

Boston ridge vent: two methods

Sheathing

Shingles

5-in. overhang

2x4 sleeper

Upper ladder

Fascia

Lower ladder

Continuous screen

Flashing

Sleepers cut from rafter stock

Fascia

5-in. overhang

Flashing

particularly when it caps a gable that forms the entry to a house. Another design consideration: your roofer will love you if you make the width of the Boston fit his shake layout.

Skylights and other obstructions—The most important consideration in cold-roof construction is to provide adequate, unrestricted air flow from eave to ridge. On a standard gable roof this is fairly simple, but when you start building cold roofs with hips and valleys and skylights, you have to be certain that the air will flow all the way to the ridge. You don't have to be a rocket scientist to figure this out, but you do have to pay attention and think about it a little.

The rough framing of skylights should be boxed in with sleepers (bottom photo, previous page). Additional sleepers above and below the skylight should stop short of the skylight framing so that air will flow around the obstruction. At hips and valleys, sleepers on adjoining roof planes should be oriented to channel air upward (top photo, previous page). By the way, don't forget to leave the skylight wells a little deeper and the plumbing vents a little taller than normal. It wouldn't be good for either one to end up inside the cold roof.

Looking for an easier way—By now you're wondering if there isn't an easier way to prevent ice dams in snow country. Some builders have tried, with uneven results, to get away with a modified cold roof. This method eliminates the sleepers and is accomplished by holding the insulation down in the rafter bays and trying to get the air to flow to the ridge between the insulation and the sheathing. I've seen cardboard baffles inserted 2 in. down from the top of 2x12 rafters, with R-30 batts underneath. This method can work, but it does sacrifice 2 in. of insulation space. The most successful use of the method I know of involved using 16-in. deep wood I-joists as rafters; there was more than enough airflow space above the insulation.

The major problem with this approach comes when condensation forms above the insulation and drips down to soak it. Even with a standard cold roof, it is very important to install a vapor barrier on the warm side of the insulation to keep moisture from migrating into the rafter bays.

One alternative that works—The need for a cold roof comes largly from a penchant for vaulted ceilings. Cold attics, whether they are built with trusses or rafters and ceiling joists, work just fine to prevent ice damming *if they get enough air flow.* This means you must provide gable-end vents of sufficient size and/or provide soffit and ridge vents (for more on roof venting, see *FHB* #61, pp. 76-80). I've heard few reports around here of leakage problems on well-insulated houses with well-vented cold attics. □

Steve Kearns is a builder in Sun Valley, Idaho. Photos by the author.

Drawing: Christopher Clapp

Dummy Rafter Tails

A useful technique for old-house remodelers and those who build alone

by Bob Syvanen

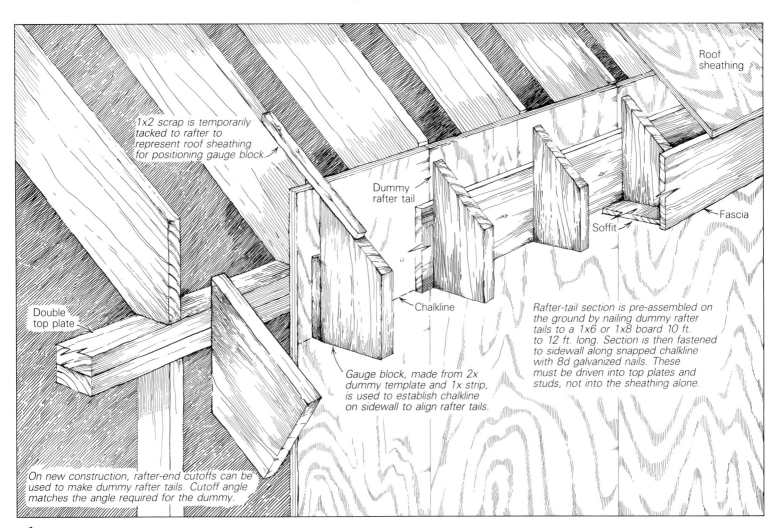

1x2 scrap is temporarily tacked to rafter to represent roof sheathing for positioning gauge block.

Roof sheathing

Dummy rafter tail

Fascia

Soffit

Chalkline

Double top plate

Gauge block, made from 2x dummy template and 1x strip, is used to establish chalkline on sidewall to align rafter tails.

Rafter-tail section is pre-assembled on the ground by nailing dummy rafter tails to a 1x6 or 1x8 board 10 ft. to 12 ft. long. Section is then fastened to sidewall along snapped chalkline with 8d galvanized nails. These must be driven into top plates and studs, not into the sheathing alone.

On new construction, rafter-end cutoffs can be used to make dummy rafter tails. Cutoff angle matches the angle required for the dummy.

A leaky roof or a backed-up gutter often leads to eaves and rafter tails that are badly deteriorated. On older houses, there's simply nothing solid to nail new fascia and soffit trim to. To deal with this problem, I cut the old rafter tails back flush with the siding, nail dummy 2x6 block rafter tails to a 1x board, and nail this in turn through the sheathing to the top plate, the cut-back rafter ends, or to both.

I've found that this dummy rafter-tail system also makes it easy for one person to nail up rafters on a new house. This procedure also eliminates the need to cut and install frieze blocks between rafters atop the plates. I extend the plywood siding 4 in. to 6 in. above the top plate, and it holds the bottom end of the rafter in place while I nail the ridge end. This sometimes lets me get away with using standard-length (20-ft. or less) rafter stock instead of having to special-order something longer. In these cases, I make up the blocks from the scrap lumber that's all over the job site. When length isn't a problem, I use cutoffs from the rafters themselves for the dummy tails. They've already got the correct angle cut at one end, so I have only to cut them to length and width.

Step by step—Not all framing material is a consistent width, so if I'm making up my dummy rafter tails from 2x6 stock, I begin by running the block stock through a table saw to make sure all the pieces are the same width. I cut as many rafter-tail blocks as there are rafters, then nail them to a 1x6 or 1x8 board. On a house 36 ft. long, I make up three 12-ft. sections for each side—longer lengths are too unwieldy.

To position the block strip, I use a gauge block made up of a rafter-tail block with a piece of 1x stock nailed to the back. I align the gauge block with the top of the rafter (on new construction, a 1x2 temporarily tacked to the rafter top to represent roof sheathing helps here), hold it against the sidewall sheathing, and mark the bottom. It's best to mark both ends of the wall and the middle, then snap a line the full length. This will show up any variation that might have resulted from the guide marks having been taken off a high or low rafter. To get a good long-line snap, I thumb the middle of the line and snap each side.

Once the dummy rafter tails are in place, I install the roof sheathing over them, tying the whole business together. The end result is rafter tails that line up straight as string. □

Bob Syvanen is a consulting editor of Fine Homebuilding *magazine.*

Attic storage with a truss. Not all trusses eliminate attic storage or attic-level living spaces. Attic trusses carry extra loads on the bottom chord and provide a room-size cavity between webs.

All About Roof Trusses

Trusses can be used to frame even complex roofs, but they must be installed and braced properly

by Stephen Smulski

Not too many years ago, you didn't find roof trusses in homes with anything more elaborate than a gable roof. Thanks to computers and sophisticated design software, that era is long gone. Name just about any roof these days—gable, hip, saltbox, mansard or gambrel—and it can be framed with engineered trusses. Precision-made from dimensional lumber and metal connectors called truss plates, prefabricated trusses have revolutionized residential roof construction over the last three decades. Roofs on more than 75% of all new houses in the United States are built with trusses instead of with conventional 2x framing, and it's not hard to understand why.

Trusses give builders a bigger bang for their buck. Truss-framed roofs can be erected more quickly and with less skilled labor than stick-built roofs. Closure against the weather is faster because trusses and roof sheathing often can go up on the same day. Trusses mean more flexible floor plans because they can span longer distances without interior bearing walls than conventionally framed roofs. Trusses are efficient in their use of lumber. Where a conventionally framed roof might require 2x8 rafters spaced 16 in. o. c., for instance, trusses for the same roof might be made entirely of 2x4s and spaced 2 ft. o. c. and use 15% to 25% less wood.

When you're used to beefy 2x rafters, trusses take a little getting used to: They look spindly. But once they are set and braced properly, trusses are stronger than stick-framed roofs. A truss is really a series of triangles, a geometric shape that is difficult to distort under load. Unlike common rafter stock, much of the material that goes into roof trusses is machine tested for strength and is held together by engineered truss plates sized for the loads they will carry (for more on how roof trusses actually work, see sidebar p. 106).

For all their advantages, though, roof trusses must be handled differently than regular 2x rafters. They have strength only in a vertical position and can't be banged around a job site like a 2x12. Proper installation techniques and bracing are critical, and trusses can't be modified in the field without radically altering their strength.

How does the cost of a trussed roof compare with a conventionally framed roof? The answer depends on the complexity and the size of the roof. But because trusses use less material, they look more attractive as the cost of framing lumber continues to rise, and high-quality framing stock becomes harder to get. The only real way to know is to price both options, but keep in mind that a conventionally framed roof takes more skill to build.

Shapes and sizes—Trusses are as varied as the houses they go on (drawings right) and can be combined to create complex roof shapes. Some trusses look similar—the king post, queen and Fink trusses, for example, all have the same shape—but can be distinguished by the signature web patterns inside. Because each web design distributes force differently, these trusses are rated for different loads and spans even though they look very much alike.

Residential roof trusses range from 15 ft. to 50 ft. long and from 5 ft. to 15 ft. high. Length and height are determined by roof pitch and span plus cantilever, if there is any. Tall trusses are sometimes made as two separate trusses so that they can be shipped over the road without hitting power lines and overpasses. Called piggyback trusses, the two parts are joined on site with plywood or metal gusset plates.

Where common trusses can't be used or aren't appropriate, special trusses fill the bill. What if, for instance, plans call for a chimney or a dormer that's wider than the normal spacing of the trusses? If you're framing a roof conventionally, you just head off one or more rafters to create the oversize hole. But trusses can't be cut. Instead, master and split truss sets are used. Truncated in midspan to form the opening, split trusses are headed off to full-span master trusses on both sides to create the opening in the roof framing.

Most roof trusses have webs that run at an angle between top and bottom chords. One exception is the gable-end truss (also called a gable-end frame in the industry) in which webs run vertically. These trusses ride atop a building's end walls and must be supported along their entire length, so they function more like a wall than a truss. Shorter than the last common truss by the depth of its top chord, a drop-top gable-end truss makes ladder-framing wide overhangs a snap.

Trusses eliminate attic storage or living spaces because webs get in the way—at least that's the oft-heard but unfounded concern. In fact, there are attic trusses that are perfect for steeper roofs and garages because they are designed with room-size central openings (photo facing page).

Combinations of several kinds of trusses are used to frame the roofs of L-, T-, H- or U-shaped houses. To eliminate the partition between the main house and the ell, girder trusses, consisting of two or three trusses fastened side by side, span the opening where the two legs of the house meet. Common trusses on the main house are clipped flush at one end and hung from metal

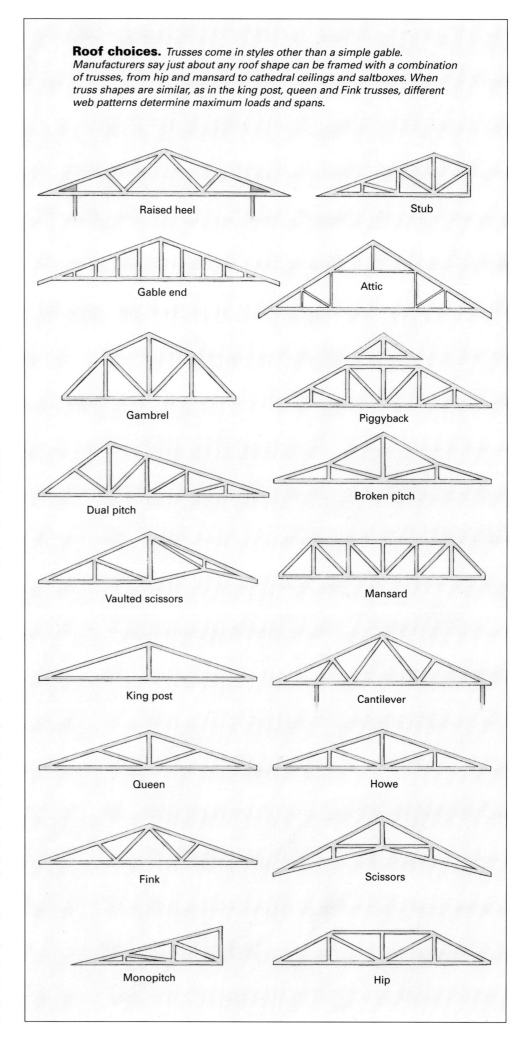

Roof choices. *Trusses come in styles other than a simple gable. Manufacturers say just about any roof shape can be framed with a combination of trusses, from hip and mansard to cathedral ceilings and saltboxes. When truss shapes are similar, as in the king post, queen and Fink trusses, different web patterns determine maximum loads and spans.*

Raised heel

Stub

Gable end

Attic

Gambrel

Piggyback

Dual pitch

Broken pitch

Vaulted scissors

Mansard

King post

Cantilever

Queen

Howe

Fink

Scissors

Monopitch

Hip

hangers on the girder truss. A series of step-down valley trusses installed on top of the common trusses of the main roof extends the ell roof back to the main roof. It's a trussed version of valley framing with conventional 2x stock. Hip roofs are framed similarly with common, girder and step-down hip and jack trusses. Scissors trusses and vaulted scissors trusses give instant cathedral ceilings. With single- and double-cantilever trusses, porches, entrance roofs and wide overhangs are simply extensions of the truss.

Common trusses are fabricated with a variable top-chord overhang and a variety of soffit-return details for box and closed cornices. And where thick ceiling insulation that extends to the outside of the top plate and an airspace above are needed, raised-heel trusses do the trick.

Specifying trusses—With so many choices, specifying trusses might seem daunting. Not so. In fact, truss manufacturers take virtually all of the sting out of ordering a roof. A basic specification starts with a list of each type of truss that will be needed (common, gable-end, etc.) and how many of each. Other information the manufacturer needs includes the span of the roof, the pitch, the top-chord overhang, the kind of end cut and soffit-return details you want, gable-end preferences and any special loading requirements (for a slate roof, for example, or HVAC equipment that is to be mounted in the attic). The specs for my 22-ft. square garage were pretty simple: 10 common trusses; two gable-end trusses; 8-in-12 pitch; 22-0-0 span; 1-ft. overhang, plumb cut, no soffit return. Many of the larger truss manufacturers and building-material sup-

At the factory.
Roof trusses are turned out quickly and precisely, thanks to computer-controlled cutoff saws and special assembly tables. At Wood Structures, Inc., in Biddeford, Maine, trusses are assembled either on huge, fixed tables (above) or on movable bases (left). When finished, trusses are moved into the yard on carts that keep trusses upright (below).

Strength in triangles

Triangles are naturally rigid geometric shapes that resist distortion. In fact, you can't change the shape of a triangle unless you change the length of one of its three legs. That's the secret to the strength of a roof truss. Regardless of its overall shape, all of its chords and webs form triangles. Stick-built roofs operate on the same principle, with rafters, ceiling joists and collar ties forming the triangles. The pieces that make up a truss work together. Cut any one of them, and the truss is compromised and weakened.

Under the weight of sheathing and roofing, a roof truss as a whole is stressed in bending. Its chords and webs, however, are stressed principally in either tension or compression (drawing right). Top chords, which are in compression, push out at the heel and down at the

peak. The bottom chord, firmly fastened to the top chords, is stretched in tension to resist outward thrust. The result is a stable, self-balancing structure.

Loads on the individual pieces that make up a truss can be dramatically different. Let's suppose the truss in the drawing has a span of 30 ft., a total rise of 7 ft. 10 in. and a pitch of 6-in-12. Under typical roof-design loads (42 lb. per sq. ft. on the top chord and 10 lb. per sq. ft. on the bottom chord), the compression force between panel points in the top chords varies between 2,400 lb. and 2,079 lb. Along the bottom chord, forces in tension run from 2,124 lb. to 1,459 lb., depending on where along the span they're measured. Contrast those forces with those working on the webs— what a difference! The two short web pieces experience 505 lb. in compression, and

the two long webs have to resist 743 lb. in tension. The relatively light forces are why web pieces don't have to be stress-rated and why materials used in top and bottom chords usually do.

One important difference between stick-built and truss-framed roofs is that ceiling joists rarely span the width

of the building. Instead, they bear on interior partitions as well as on exterior walls. Trusses are almost always designed to bear only on exterior walls, with the webs connecting the top and bottom chords, providing intermediate support. That opens up many more possibilities for floor plans. —S. S.

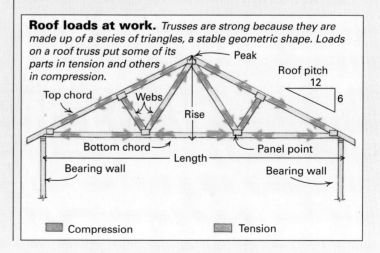

Roof loads at work. *Trusses are strong because they are made up of a series of triangles, a stable geometric shape. Loads on a roof truss put some of its parts in tension and others in compression.*

Peak
Roof pitch
12
6
Top chord
Webs
Rise
Bottom chord
Panel point
Length
Bearing wall
Bearing wall

☐ Compression   ☐ Tension

pliers carry stock trusses in sizes and styles most popular in their areas. With no production lead time, you can order stock trusses on Monday and have the roof closed in by Friday.

Specifying trusses for complex roofs is easier, not harder. All you really have to do is take the framing plan to your building-materials supplier or to one of the nation's 1,500 or so truss fabricators, and they'll do the takeoff for you. Most truss manufacturers purchase the machinery, the plates and the engineering services of one of about eight connector-plate manufacturers. Plate makers actually engineer and design the trusses, and truss manufacturers assemble and sell them. Some truss manufacturers have in-house engineers and design software.

Design and fabrication—Once the specs are known, trusses are designed by a computer, with building-code required roof, ceiling, wind and snow loads, as well as any special loading conditions, taken into account. An engineering drawing details the forces that develop in each chord and web under the design loads. Engineering specs also include lumber species, size and grade for each chord and web; gauge, size and orientation of each connector plate; truss dimensions and pitch; and the location of permanent bracing. Engineering drawings are supplied by the truss manufacturer, who passes them on to the contractor. If you're responsible for setting the trusses, and you didn't get a copy of the drawings, be sure to ask for them.

Integrity of the truss depends on the integrity of its metal plate connectors. Stamped from 16-, 18- and 20-ga. structural steel coated with zinc, plates have many integral teeth $\frac{5}{16}$ in. to $\frac{9}{16}$ in. long. There are about eight teeth per sq. in., and plates are sized according to the level of stress they have to transfer between members.

Trusses are made mostly from southern pine, Douglas fir and the woods of the spruce-pine-fir group. That includes eastern and Sitka spruce, lodgepole, red and jack pine and western and balsam fir. Truss manufacturers start by cutting 2x members so that they are the right length and have the correct angles at the ends. Some factories use computer-driven saws that can change settings in about half a minute and produce multiple cuts very rapidly and precisely. The kind, the size and the grade of lumber for each chord and web on the cutting list is based on how great a force each must resist while under load. Highly stressed as a rule, chords are usually made of lumber that has been stress-tested to ensure performance. Webs, because they are usually subjected to lower stresses, are more likely to be #2, #3 or even stud grade.

At the factory—There are two common methods for building trusses in a factory (photos facing page). In one, pedestals with electromagnetic bases are arranged in the shape of the truss on a steel floor, with one pedestal at each panel point. Once all of the pedestals are in place, chord and web pieces are laid out, their ends tightly butted, and then clamped. Connector plates are positioned on both faces of the joint, and a hydraulic C-clamp suspended from a gantry squeezes the teeth of both plates into the wood simultaneously.

Another way of making trusses is to assemble them on huge metal or wood tables. The tables are drilled to accept a series of pins and clamp fixtures that hold the truss pieces in place. Chords and webs are placed in the jig, then panel points are lifted, and connector plates are slipped underneath. Another plate is set on the exposed face at each panel point. Both plates are pressed into the wood at once by a mechanized roller that travels the length of the table.

It may take a couple of experienced workers 30 or 40 minutes to lay out pieces for a new truss on one of these tables. But once the pins and the clamps have been adjusted correctly, workers can put their measuring tapes away; all they have to do is pick up precut pieces of stock, put them in the jig and add the connector plates.

To complete the process, and to assist builders in setting the trusses, most truss manufacturers affix brightly colored "Caution!," "Warning!" and "Danger!" tags at critical locations such as cantilever bearing points and permanent lateral bracing sites. Don't ignore them.

Completed trusses are stacked, banded and stored in the truss yard, either vertically or horizontally. When stored on their sides, trusses are elevated off the ground on stringers spaced to minimize lateral bending.

Delivery, handling and erecting—Trusses are hauled by truck, with trusses either lying on their sides or cradled vertically (often upside down) in a special trailer—trucks can haul up to a half-dozen roofs at the same time. Ideally, trusses are unloaded at the job site with a forklift or a crane, but most are gingerly dumped on level ground. Trusses should always be elevated off the ground and protected from the weather under a loosely draped tarp. When unloaded vertically, the bundle of trusses that rests on its top chords should be braced on both sides to prevent it from falling over and to keep trusses from toppling when the band is broken.

Depending on the truss span and the height of the building, trusses are erected either by hand or by crane, and occasionally by forklift. With one-story buildings, trusses under 30 ft. usually can be raised manually, but longer trusses should be hoisted by crane. A crane is a must for buildings over one story (photos p. 108), regardless of truss length. Whether carried or hoisted, trusses should always be held vertically when moved. When held horizontally, lateral flexing and bouncing can overstress the connections in the truss, causing plates to loosen or pop out. Long trusses are especially vulnerable to this problem.

Installation starts with a gable-end truss. Setting trusses by hand on a one-story building might go like this: With its peak pointing down, a gable-end truss is carried into the house, and its ends are positioned carefully on top of the sidewalls (top photo, p. 109). Then, with a worker at each sidewall, the rest of the crew uses Y-shaped poles to rotate the truss until it's upright (middle photo, p. 109). To prevent damage during lifting, two poles should be used. Each pole should be positioned at the panel point closest to the quarter points of the span. If only one pole is used, it's placed at the peak. After carpenters make sure the overhang is correct, the bottom chord is toenailed to the end-wall top plate with 16d nails (bottom photo, p. 109).

Once the gable-end truss is in place, it must be braced to the ground. Common trusses are then raised sequentially in the same way. Each should be secured in place with temporary lateral bracing that goes back to the gable end (more on this later). That's why it's essential that the gable-end truss be securely braced to the ground—all other trusses will be braced against it. It's also very important that the 2-ft. o. c. spacing be maintained at the heel and the peak of each truss and that each goes up square and plumb.

Setting trusses with a crane—Good rigging practice prevents damage when setting trusses by crane. No truss should ever be lifted by its webs. Trusses up to 20 ft. long can usually be lifted with a cable looped around the top chord at midspan (drawing p. 108). A tag line, which doesn't support weight but gives you a way of steering the truss when aloft, is lashed to one heel and is used to guide the truss into position. Trusses up to 40 ft. long are typically hoisted at two symmetrical lifting points separated by half the span. Again, cable ends are secured around the top chord. A tag line is needed as well. Lifting 40-ft. to 50-ft. long trusses without lateral flexing generally requires a spreader bar with three cables. Typically one-half to two-thirds of the truss' length, the bar is centered over the truss. Cable ends looped around the top chord should point in slightly. Tag lines attached to both heels increase control.

Once in place, trusses are customarily toenailed to the top of the wall with 16d nails through slots in the heel plates. While adequate in most instances, toenailed fasteners can withdraw under the uplift forces exerted by high

For more information

The Truss Plate Institute
583 D'Onofrio Dr.
Suite 200
Madison, Wis. 53719
(608) 833-5900
(Ask for HIB-91, *Commentary and Recommendations for Handling, Installing and Bracing Metal Plate Connected Trusses*, $7.)

Wood Truss Council of America
5937 Meadowood Dr.
Suite 14
Madison, Wis. 53711-4125
(608) 274-4849
(Ask for *The Metal Plate Connected Wood Truss Handbook*, $39.95.)

Lifting trusses by crane. *To set larger trusses, and any truss on the second floor, a crane will be required. For spans up to 20 ft., a single line from the peak works fine. For spans up to 40 ft. long, two chokers should be used, each at a quarter point on the top chord. For trusses up to 60 ft., a spreader bar is a good idea. Three chokers, the two outside lines slightly pointed in, should be attached. In all cases, a tag line helps control the truss.*

Going up? Get a crane.

Large trusses, and any roof truss headed for a second-floor installation, go up by crane. This piggyback truss (above) is being lifted by two chokers with a tag line for control. Pallets and 2xs beneath the stacked trusses on the ground help prevent bending. With a gable-end truss in place (below), the rest of the roof quickly follows. The gable-end truss on this roof has been braced securely, and trusses installed subsequently are tied to the gable-end truss to prevent the trusses from tipping over.

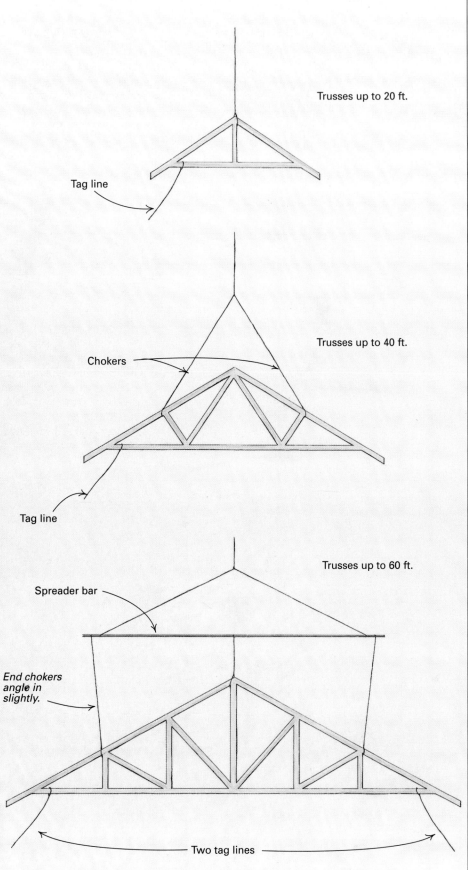

Trusses up to 20 ft.

Tag line

Trusses up to 40 ft.

Chokers

Tag line

Trusses up to 60 ft.

Spreader bar

End chokers angle in slightly.

Two tag lines

winds. This became evident in the aftermath of Hurricane Andrew in south Florida. If you want uplift resistance, you've got to use metal framing anchors or straps for truss-to-wall connections. Scissors trusses are a little different. Because these trusses have a significant horizontal thrust by nature, one heel must be free to move so that the truss won't push walls out of plumb when it's under load. The solution is a framing anchor with a horizontal slot (for more on framing anchors, see *FHB* #43, pp. 44-49).

Trusses should never be attached rigidly to interior partitions because this can induce bending forces that trusses aren't designed to carry. When trusses are nailed directly to interior partitions, cracks can open at wall-to-ceiling junctions, or partitions might be lifted right off the floor because of something called truss rise (for more on truss rise, see *FHB* #81, pp. 54-59).

Temporary bracing prevents collapse—After the heels have been nailed, the top chord of the truss must be secured by temporary lateral bracing. Starting at the heel and working up to the peak, 2x4 bracing is usually installed along the top chords at about 8-ft. intervals, with the correct interval depending on the truss span. Bracing should span four or five trusses and should be fastened to each truss with two 16d nails. The ends of the 2x braces should overlap at least two trusses. Bottom chords need to be braced, too, at intervals of about 10 ft. across the span. Although I've never seen it in use, there is a product called the Truslock (Truslock, Inc., Rt. 1, Box 135, Calvert City, Ky. 42029; 800-334-9689) that's designed to brace trusses quickly while maintaining the correct spacing. The Truslock is a folding metal brace that locks over the top chords of trusses as they are set. Even with the Truslock in place, proper bracing of gable ends and bottom chords is still necessary.

While helpful for maintaining on-center spacing, lateral bracing won't prevent connected trusses from tipping over as a unit, just like a row of dominoes. To prevent this catastrophe, trusses must be braced diagonally, either across the top chords or through the webs. With the first option, bracing is laid at a 45° angle across several trusses on both sides of the peak. A row of this diagonal bracing should start about every 30 ft. as you walk the ridge, so most houses will get only one

For small trusses, poles will do. First-floor trusses with relatively short spans can be poled up. Supported by two outside walls (top), trusses are rolled upright with a Y-shaped pole (middle). When the truss is plumb, it must be braced to other trusses and toenailed in place. In this installation, the use of two poles would have been better.

or two lines of diagonal bracing. When run through the webs, bracing starts beneath the top chord against the web closest to the center of the gable-end truss. Descending at a 45° angle, the bracing should cross several trusses, nailed to each web as it passes, and end at a web just above the bottom chord. Through-the-web diagonal bracing is often left in place.

Bottom chords are also braced diagonally in each corner of the building. In all cases, 2x4 bracing is fastened with two 16d nails to every truss it passes. Like those found in the brochure truss manufacturers provide to the contractor with every shipment of trusses, these guidelines are based on the recommendations of the Truss Plate Institute (see For more information, p. 107).

Don't ignore them. Inadequate temporary bracing is the chief cause of truss collapse during erection.

Closing in—With the trusses in place, top-chord temporary bracing is removed, truss by truss, as sheathing is laid. Ideally, each panel is fully nailed with the proper-size fastener at the recommended spacing before moving on; a panel tacked in place with a few nails may not provide the same resistance to lateral movement as the just-removed bracing. With a crane on site, you may be tempted to hoist all of the sheathing to the roof at once. Don't do it. Trusses can buckle, be damaged or broken under the concentrated load exerted by such heavy weights.

The size, the location and the methods of attachment of permanent bracing is the responsibility of the building's architect, designer or engineer. Permanent bracing works with the building's other structural elements to achieve structural integrity. Top chords of trusses are assumed to be permanently braced by the roof sheathing. Long webs and bottom chords not braced by a rigid ceiling, as in a garage or over a suspended ceiling, for example, may need to be permanently braced to prevent lateral buckling. In residential construction, permanent diagonal bracing running through the webs isn't usually necessary. Plywood sheathing provides enough resistance to racking when combined with bracing nailed to bottom chords in the corners. According to the Truss Plate Institute, building codes around the country require that roof trusses be braced, leaving detailed instructions to the truss manufacturer or the designer of the building. In some areas, like hurricane-prone south Florida, local codes may have special requirements for permanent bracing.

Damaged trusses—What should you do about the odd truss with a broken web or a popped plate or two? The temptation is to pound the plate back in or sister a 2x over the break. But the right thing to do is to contact the truss manufacturer for advice. Why? First, once a truss is damaged, it no longer acts like a truss. Second, whoever does the repair assumes responsibility. And in today's litigious society, that's no small matter. For the same reasons, never cut, notch, drill or modify a truss without first seeking engineering advice. More likely than not, the truss engineer will come up with a workable repair scheme. □

Stephen Smulski is a consultant in wood performance problems in Shutesbury, Mass., who also inspects truss manufacturers on behalf of The Truss Plate Institute. Photos by Scott Gibson except where noted.

Truss Frame Construction

A simple building method especially suited to the owner-builder

by Mark White

Standard frame construction is complex and can be baffling to the first-time owner-builder. When the purposes and natures of foundations, sills, sill plates, floor joists, partition walls, studs, cripples, headers, ceiling joists, top plates, rafters and sheathing are taken one at a time they can be understood. But novices have a hard time handling the complexity once they are staring at all the pieces on their sites.

I have been teaching building on the college level for the past six years, trying to find a method that would reduce house construction to simple elements. Having tried balloon framing, platform framing, tilt-up walls, post and beam, and variations of them all, I now think I've found an answer: truss frame construction.

The truss frame is not new. Contractors, and individuals all over the world, have fiddled with the idea for years. It began to attract more attention in the United States when the Department of Agriculture's Forest Products Laboratory in

Madison, Wis., erected an experimental building that combined floor, walls and roof in single truss sections. I was immediately taken with the simplicity of the concept. It looked like a system that would enable the owner-builder to come up with a sound, useful structure on the first attempt, given some basic training and guidelines.

To test this building concept, I sketched up a set of plans and ordered the appropriate lumber from one of our local sawmills. The lumber was ready in January. In February, I began work on the foundation, and I set the sill timbers in early March. The building, which I planned to build alone to see how well the system would work, was to be used as a rental unit. It would have 12-in. floors, 10-in. walls and a 12-in. roof, all stuffed with a nominal 12 in. of insulation, for an insulation value of R-45.

Our climate is quite mild, as Kodiak, an Alaskan island in the north Pacific, is warmed by the Japanese current. Winter temperatures rarely

drop to 20°F, and average between 30°F and 40°F most of the time. Still, the winters are long and our primary heating fuel is oil, which is delivered by tanker from the lower forty-eight. The price of oil hasn't gone down in years, which leads us to think hard about proper insulation.

Foundations—Concrete costs a lot up here (about $165 a cubic yard), so many foundations are either creosote posts or treated wood. I opted for posts, because this is the fastest method and disturbs the soil the least.

The frost line here is 6 in., and bedrock is usually between 18 in. and 36 in. below the surface of the soil. I dug into glacial till—a mixture of hard clay and shale gravel a few inches above the actual bedrock.

The posts are Douglas fir, pressure-treated with creosote and about 12 in. in diameter. The ends that go into the holes are cut off squarely, then covered with another coat or two of creo-

The author built his first truss frame house alone to test the simplicity of the technique. At left, he winches the completed trusses onto sills set atop a post foundation. Center, a framed and sheathed partial truss is tilted into place as an end wall. It will be toenailed and temporarily braced. Right, most of the trusses are

sote to protect the center where the pressure treatment has failed to reach. I then nail a few pieces of heavy asphalt shingles (smooth side in) over that end to keep water from wicking up through the center of the post. When the post is in the hole, an eventual burden of 6,000 lb. to 10,000 lb. of house forces the asphalt into the wood fibers of the end grain and pretty much seals the pores. It takes 20 or 30 years to rot out the untreated center of a Douglas fir post, but I've seen it happen. The houses we build should last at least 200 years—for this reason I'm interested in having the posts last that long as well.

We carefully dig holes by hand into the 6-in. layer of glacial till and pour about a gallon of clean dry sand into the bottom of the hole instead of using a concrete pad. The sand is easier to work with if the posts need to be shifted to line up properly, and we have had little evidence of settling. We wrap the part of each post that is going to be in the ground with a few layers of 6-mil polyethylene to reduce the leaching of poisonous creosote into the groundwater. The plastic wrapping would deteriorate rapidly in sunlight, but it lasts a long time underground.

After the plastic goes on, the post is dumped into its hole, rotated a half turn in the sand and then propped into correct alignment with a few wedges jammed into the hole on one side or another. Once all the posts are in position, they are aligned with a transit, and everything is tied together with rough-cut 2x6s fastened with hot-dipped, galvanized, 20d nails. Extensive cross-bracing is installed before any weight is placed on the piles. The bracing is extremely important, because the soil is so shallow that it lends little racking resistance to the system. Once the bracing is in, the holes can be filled and tamped around each post. Sand makes the best fill, but we usually wind up using the dirt that came out of the hole.

We usually space posts 6 ft. o.c., forming "strings" of them to support two 8x12 sill timbers along the length of the building. Near each end, we reduce the spacing to 3 ft. or 4 ft. o.c. to support the greater weight of the end walls and the extra load transmitted to them by the roof overhangs. We space the parallel strings of posts between 14 ft. and 20 ft. apart, depending on the carrying capacity of the floor joists the sills are supporting. Post foundations on sand and bedrock work well as long as the quality and spacing of the individual members is kept within reason, and the cross-bracing is adequate. We tend to be conservative, planning shorter spacing than the maximum indicated by charts and tables. A foundation is not worth skimping on. Besides, our cost only runs between $200 and $300 per structure—dirt cheap compared to the cost and labor associated with concrete.

The first house—Teaching duties and a building project in a remote village kept me from further work on the truss house until May. Then I cut out and assembled a single truss on the sill timbers. I used it to pull master patterns, from which I then traced the necessary shapes on 10-in. and 12-in. rough-cut spruce planks. I used a portable circular saw and a small chainsaw to cut out the pieces for the rest of the trusses.

The Forest Products Lab's original truss frame design called for the use of standard 2x4 material for all chords and webs and results in many more pieces in each truss. I stayed with 1½-in. by 10-in. and 1½-in. by 12-in. material in the interest of simplicity. My trial structure was to be 20 ft. wide by 24 ft. long, with outward-sloping walls and generous porches all around. The outward-sloping walls were an experiment aimed at providing more visual interior room for a given area of floor space. They did provide the room, but for a few dollars more the side walls near the floor could have been kicked out a bit under the same roof and I would have had even more room. In a word, the experiment was successful, but I wouldn't repeat it. A frame spacing of 24 in. called for 11 full trusses, two partials for the end walls, and a total of four roof trusses to support the porch overhang at the ends.

A partial truss is a truss that has gussets on one side and studs on the other side. Trusses are used on the end walls only to define the shape and outline of those walls and to hold the studs in that configuration before the walls are tilted up into position.

Roof trusses are made up of rafters and collar ties, without vertical members. If the overhang at the gable end is less than 3 ft., it is possible to use roof trusses supported only by a sturdy 2x12 fascia board nailed to the other rafters along the eaves. If the overhang is greater than 3 ft., some arrangement of conventional headers and cripples is necessary. This often makes even the cross tie unnecessary.

Construction of the trusses went according to plan, but including a floor joist as a part of each unit turned out to cause more trouble than it was worth. It meant that there was no floor to work

erected, toenailed to the sills, and stabilized by the plywood nailed along their sides. In this first house. floor joists were a part of each truss. This resulted in great strength and stability, but made construction a bit awkward, since there could be no platform to work on until all the trusses were in place.

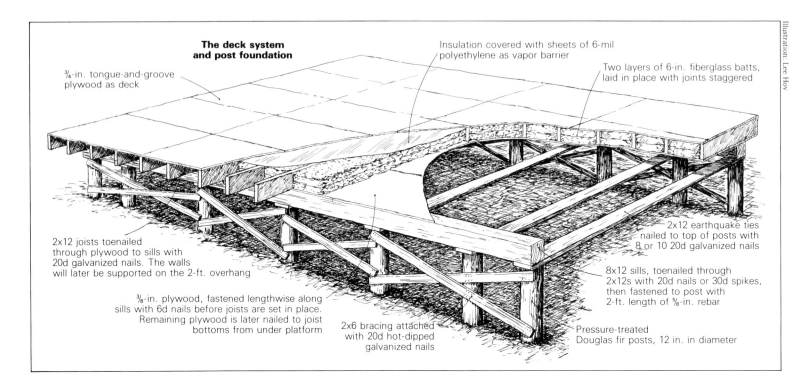

The deck system and post foundation

¾-in. tongue-and-groove plywood as deck

Insulation covered with sheets of 6-mil polyethylene as vapor barrier

Two layers of 6-in. fiberglass batts, laid in place with joints staggered

2x12 joists toenailed through plywood to sills with 20d galvanized nails. The walls will later be supported on the 2-ft. overhang

⅜-in. plywood, fastened lengthwise along sills with 6d nails before joists are set in place. Remaining plywood is later nailed to joist bottoms from under platform

2x6 bracing attached with 20d hot-dipped galvanized nails

2x12 earthquake ties nailed to top of posts with 8 or 10 20d galvanized nails

8x12 sills, toenailed through 2x12s with 20d nails or 30d spikes, then fastened to post with 2-ft. length of ⅝-in. rebar

Pressure-treated Douglas fir posts, 12 in. in diameter

Illustration: Lee Hov

on, so moving materials, assembling them and erecting them was awkward work.

The building did go together with less effort than one built with either the standing stick or the tilt-up wall method. And the truss assembly with its plywood gussets made an extremely strong and rigid structure. If anything is wind and earthquake-proof, it is this building. I was heartened by the progress and by the rigidity of the building, but felt the system could be simplified even further.

An improved design—The following fall, I designed a 20-ft. by 26-ft. house with straight walls 6¼ ft. high. This time the floor would be built and insulated as a separate unit (drawing, above), with the trusses and end walls erected on top of it. This is much easier, and almost as strong.

It took an inexperienced seven-person crew (my class) six hours to build and insulate the floor. First, we laid our joists (in this case, 20-ft. long 2x12s) over two sill timbers, spacing them 24 in. on center. Then we nailed ⅜-in. CDX ply-

wood to the joists from underneath and dropped in two layers of 6-in. fiberglass insulation, making sure to stagger their joints. On top of this went a 6-mil vapor barrier, then either a plank or a plywood floor. Dropping insulation in from above is an easy 10-minute task. If we had had to install it from below, it would have become a frustrating, eye-irritating chore.

Each truss designed for this building consists of five main pieces: two wall studs, two rafters and one collar tie. The class cut and assembled the required twelve full trusses, two end walls and three roof trusses in exactly 14 hours of hard labor. To eliminate inaccuracies, each truss was assembled on top of the master truss. All of the ⅝-in. plywood gussets were nailed with great quantities of #6 galvanized nails.

I have used gussets of ½-in. plywood, but they sometimes break when they're fastened to only one side and the truss is being flipped over to have the rest nailed onto the other. Using ⅝-in. plywood has eliminated the problem. I would recommend ¾-in. plywood for gussets on larger trusses and in two-story houses. Once gussets are nailed onto both sides of a truss, breakage is extremely unlikely, even with ½-in. plywood.

Truss members are very thick to provide room for lots of insulation. The trusses are thus massively overbuilt, so a builder need not worry much about a structural failure. A neophyte designing a truss pattern need only include a substantial enough collar tie to keep the rafters from spreading out near their bottoms. The collar ties should equal about half the span of the rafters. We usually use 2x12 ties, which can safely span 16 to 18 ft. On longer spans, it's safer and cheaper to use smaller dimensioned lumber, with a center tie fastened between the center of the collar tie and the peak of the roof. The center tie adds strength and keeps the ceiling from sagging and the long collar ties from warping.

We scheduled raising the frame for a Saturday so the students could work all day. Moving the sections and clearing ice from the floor's plat-

The completed first house. Sloping side walls increase the volume with limited floor space.

form took about two hours. Putting up the end walls and the trusses themselves took 17 minutes. Passers-by were amazed at the speed with which the building took shape.

Initially, the bottoms of the studs were toenailed to the floor, with each stud positioned over a floor joist. The first course of plywood sheathing was nailed to the floor perimeter and to the studs as the trusses were raised. This course secures the studs to the deck and keeps the trusses erect. More plywood sheathing on the walls and 1x6 decking on the roof locks everything together.

Alignment of the walls and roof was perfect. This turned out to be the straightest and squarest building I've ever seen assembled—thanks to a carefully leveled floor and to the uniformity of the trusses.

Insulation—The class met for four hours one final Saturday to get the building weathered in. To insulate the roof, we strung nylon twine under the rafters and stapled it in place to hold the fiberglass. In about 15 minutes, two layers of 6-in. insulating batts were laid in from above. We then decked over the roof with 1x6 spruce, trimmed and covered it with a layer of 55-lb. roofer's base felt. This layer will eventually be covered with asphalt shingles, but it will protect the building until good weather returns.

We have built a number of buildings with floor, wall and roof sections insulated to R-45. They are effective, allowing us to heat the average home with 15 to 20 gallons of fuel a month during the winter. A conventionally insulated house of equivalent size may gobble up to 350 gallons of fuel a month during a similar heating season.

In most of these buildings, I used Owens Corning 6-in. fiberglass in the roof. It seems to come 5½ in. thick, so two layers gave me 11 in. of insulation and a 1-in. ventilation space. I used only friction-fit batts or rolls, no foil face. In the floors and walls, I used Johns-Manville 6-in. fiberglass, two layers of which usually measure out to 13 in. (Owens Corning and Johns-Manville claim the same R-value, and they cost the same.) Recently I have switched to 12-in. wall cavities to make better use of the Johns-Manville insulation.

Plumbing, electrical and heating—The structures my classes and I have built appear deceptively simple in shape, form and function. They are not. Their cross-sectional designs have been carefully worked out to provide maximum floor space and volume with minimum exposed surface area. There are a number of deviations from typical construction practice that could be imposed upon any building method, but seldom are. The very nature of the truss frame structure, along with our floor-building technique, simplifies and encourages their use.

First, the floor is an almost totally sealed unit. Its vapor barrier is penetrated only by a carefully installed waste pipe and a water supply line. We use 6-in. or 8-in. interior plumbing walls, and try to back the kitchen up to the bathroom, utility room and wash room to get all the plumbing into one area.

Under the house, I make no attempt to insulate the 3-in. waste drain that flows to the sep-

Trusses for the second house did not include floor joists. A deck was built first so that work could progress easily, then—in just 17 minutes—trusses were tilted up, toenailed and temporarily braced, before being firmly connected to the deck by the first course of horizontal plywood sheathing. These workers are stringing twine beneath the rafters to hold 12-in. batts of fiberglass insulation.

tic system. It is merely angled properly and enters the ground quickly. None of the many that have been installed this way in our location has ever frozen. I usually wrap the 1-in. PVC water supply line with a short length of thermostatically controlled heat tape extending just below the frost line, and then insulate it. A neon indicator light on the upper end of the tape tells you whether it's working or not.

In Alaska, we install this 1-in. pipe inside a 2-in. pipe, which in turn should be insulated to beneath the frost line with heavy, black neoprene foam made for the purpose.

Electrical power enters the house by way of a single piece of 1½-in. conduit passing from a meter base on an outside wall to a single service panel on the wall inside the vapor barrier. All circuits emanate from this service panel, passing throughout the building in a single channel cut into the interior face of the outer wall studs. (There are two such channels cut into the faces of the studs in a two-story dwelling.) We use either Romex cable or conduit, ½ in. or ¾ in. in diameter. All cables and conduits are inside the vapor barrier, so there are no leaks through the membrane. Switches and electrical outlets don't let cold air get into the room.

The heavy insulation and complete vapor barrier eliminate drafts, convective air currents, and excessive heat losses, so there is little need in our houses for complex heat distribution systems. For heat sources we have tried woodstoves but they typically put out too much heat. We have settled on either a standard oil-fired, hot-air furnace, or an oil hot-water heater with a short loop of pipe run from the water tank to heat the air. In fact, we have a real problem finding an appropriately sized heat source. Right now the smallest furnace available is in the neighborhood of 85,000 Btu. What we really need is one that kicks out 12,000 Btu or less.

Potential uses—The frame truss system lends itself to simple buildings with repetitive sections, and I've used it to build structures with conventional shapes. But different applications and shapes are possible, because the basic truss is highly adaptable. I once built a strong and very lightweight building out of 1½-in. by 1½-in. stock and plywood glued and nailed in place. If you live in an area where labor is considerably cheaper than material you might try a truss composed of 2x2s or 2x3s and light plywood gussets glued and stapled in place to form an intricate webwork. This would eliminate the problem of direct heat transfer through solid joists, studs and rafters. It would, however, introduce the new problems of insulating between the webwork and of sealing off passages to fire and rodents. In our area rough-cut lumber is available at a reasonable price in lengths of up to 24 ft. so trusses of solid lumber are more cost effective than lighter, more intricate designs.

The use of a well-ventilated, heavy truss system instead of concrete in an underground structure merits consideration. The strength of a properly designed truss makes it a good choice in the high load conditions found beneath the earth. Another area of application would be in passive solar designs where the shape would be in one of the many asymmetric configurations, with the tall open side facing south.

The beauty of the truss system is that only one of the many frames that go into the building needs to be laid out with great care. Once that first unit is formed it's an easy and repetitive task to construct the rest of the units, using the first as a pattern to get the rest right. Raising the frame is then a simple matter involving a minimum of fiddling and measurement. □

Mark White teaches at the University of Alaska at Kodiak.

Raising Roof Trusses

Careful preparation and wise use of a crane
can allow a good-size roof to be framed
and sheathed in a day

by Rick Arnold and Mike Guertin

We learned truss raising the hard way. Fingers trapped between sliding trusses were broken, and backs were strained muscling 32-ft. wide trusses up two stories for steep-pitched colonial roofs. The process involved six or seven guys and usually lasted a whole frustrating, tiring day. The process was tough on the trusses as well. Lifting them flat would cause the truss plates to bend or pop, ruining the truss. If we were lucky enough to get all of the trusses installed in a day,

we then faced the arduous task of lugging up all of the sheathing and roofing materials. After all of that hauling and lifting, there wasn't much enthusiasm or energy for swinging a hammer. Our lives got much simpler after we opened the phone book to the listing for "Cranes."

We have come to depend on a crane to lift the trusses for all but the smallest single-story roofs. In this article, we'll discuss how we use a crane to assemble simple gable roofs. In the next article

(see pp. 120-125), we'll explain how we build hip and valley roofs with trusses.

Before the roofing materials and trusses are dropped at the site, we anticipate where the crane will be positioned so that all materials are out of the way but still easily within reach. Before the trusses are lifted to the roof, the crane hoists the sheathing, the shingles and the interior-framing materials for the second floor (sidebar p. 119). The crane we hire can usually reach an entire

Instant, ready-made gable. With the sheathing, shingles and trim boards already attached, the only thing this gable needs is the vent. The crew will leave the peak of this gable just tacked in place until the overall length of the house is checked at the eaves and the peak.

Stacked for easy layout. With the trusses stacked neatly on the ground, layout lines are drawn on all of them at once. The line closest to the tail will be used to position the truss on top of the wall plates. The other lines are for ceiling strapping. The top chords are laid out for roof sheathing.

house and garage from one position near the outside middle of the building. We also keep the job site as clean as possible. Scrap lumber and debris can be accidents waiting to happen on raising day. Our scrap pile is always located out of the way but still within tossing distance of the house.

When the trusses arrive, we land them on top of 2x blocks as they come off the truck to keep them as flat and as much out of the dirt as possible. Sometimes trusses arrive as much as a week before they can be installed. If the trusses are not kept flat, they can warp, and after they've warped, they're much harder to install.

Layouts are done with the trusses on the ground—We mark three different sets of layout lines on the trusses before they ever get off the ground (top photo). The first set is for positioning the trusses on the walls. The second is for the strapping or furring strips to which we'll screw the drywall or plasterboard ceilings, and the last set of lines is for the roof sheathing. We also mark the front of the trusses near the tail of the top chord to avoid getting them spun in reverse when they're lifted. Before we mark these lines, the trusses need to be aligned on top of one another as closely as possible. We arrange them in a stack by tapping them back and forth with a sledgehammer until all of the bottom chords and ridges are in line. When the trusses are in a straight stack, we can make our layout lines.

In the past we aligned the trusses after they were raised. We would run a string from one gable peak to the other and bang the trusses back and forth with a sledgehammer until they lined up with the string. But all of that banging knocked the walls out of plumb, and the whole process took a lot of time. We also tried measuring in from the tail cuts to position the trusses but soon discovered that those distances varied significantly and that the peaks of the trusses didn't line up. Now we mark the wall position on the bottom chord of the trusses before we lift them; it's the quickest, most accurate way of ensuring that the trusses are installed in a straight line.

First we measure the overall width of the house where the trusses will sit. Then we locate the exact center of the bottom chord on the top truss of

Nails for attaching the gable truss are started on the ground. To facilitate setting the gable truss, nails are started along the bottom chord. When the truss is in position, a crew member on a ladder will drive the nails into the top plate of the gable-end wall.

our pile and measure back half of the width of the building minus the thickness of the wall from that point. For example, if the house is 30 ft. wide with 2x4 walls, we would measure back 14 ft. 8⅜ in. from the center point (figuring the walls at 3⅝ in.). We mark each end of the top truss and then repeat the whole process on the lowest truss in the pile. We either snap a chalkline or scribe a line along a straightedge between the top and bottom marks to transfer the layout to the other trusses. We extend our position lines slightly onto the face of the bottom chord to make it easier to see the lines when the trusses are being dropped into place.

Ceilings in our region are typically strapped with 1x3 furring strips before the interior walls and wallboard are installed. Furring strips serve several functions, one of which is adding a structural element to the trusses. If we didn't use strapping, the truss engineer would require us to run three or four 2x braces the length of the building to tie the trusses together and also to keep the gable walls straight.

We mark the lines for the furring strips 16 in. o. c. from the positioning marks at one end of the truss. Again, the same marks are made on the top and bottom trusses, and lines are drawn or snapped between them. We run a lumber crayon up one side of each line to indicate which side of the line our furring gets nailed to once the trusses are set.

The last set of layout lines is for the roof sheathing. We usually fill in any ripped sheets at the bottom of the roof rather than at the top, so we begin our layout from the ridge. First we measure down along the top chord of the upper truss from the peak to the 4-ft. increment nearest the tail of the truss. We add ⅛ in. per sheet for the H-clips plus an extra ¼ in. to our overall measurement to be safe. For example, if the top chord of the truss is 17 ft. 6 in., we would measure 16 ft. ¾ in. (4 sheets x ⅛ in. = ½ in.; ½ in. + ¼ in. = ¾ in.) If a ridge vent is being installed, we increase our extra amount to a full 1 in., and our layout line is at 16 ft. 1½ in. The top chords of the lowest truss in the pile are marked at the same points, and lines are drawn or snapped between the marks.

Gable trusses are sheathed and sided before they go up—Another big time-saver that we've found is sheathing, siding and trimming the gable-end trusses while they are still on the ground (bottom photo, p. 115). It saves us the hassle of setting up tall staging to work up high and requires just a little planning. Because the gable trusses sit on top of walls, we make sure that there are no humps in the wall or dips in the bot-

tom chord of the truss. We check the walls by eye and string the bottom chord of the gable trusses. If necessary, we rip a little off the bottom of the bottom chord of the truss.

We always plan to extend our gable-truss sheathing down beyond the bottom chord of the truss a couple of inches. To set this overhang, we run a string between the seat positions of the truss and measure from the string to set our sheathing. We always leave the sheathing ¼ in. short to avoid any binding between the gable-wall sheathing and the truss sheathing when we're setting the truss. The gable truss must be lying absolutely flat when the sheathing is applied, or it may be impossible to straighten the truss when it's installed. Any sheathing that extends above the top chord gets trimmed off flush.

If no rake overhangs are called for, siding the gable goes quickly. We start by figuring out the approximate height of the top of the second-floor windows (they don't get installed until after the trusses) and determine the siding courses from there. It's best to start one or two courses up from the bottom of the truss sheathing so that the siding doesn't interfere with the truss installation. Wood siding allows us to adjust our course exposure slightly to blend the siding on the walls with the truss siding. We've never tried this technique with vinyl siding and suspect that it would

The front of the truss is left loose. With the trusses secured at the rear of the house and the peaks locked together, the front of each truss is allowed to float until after all of the trusses are in place. The walls can then be re-strung and straightened if necessary before the fronts are nailed permanently.

Spacers lock the peaks in position. One crew member holds the trusses in place while another attaches truss spacers (made by Truslock Inc.) along the peaks of the trusses. Strapping replaces the spacers so that only a couple of 16-ft. lengths of spacers are needed. The crew member at the rear of the house taps the truss into position on the wall plate and nails it home.

be difficult to match the courses exactly. We're careful to nail the bottom course of the truss siding up high so that the course coming up from below can be slipped in later.

With no rake overhangs, we run the siding over the top chord of the truss but keep the nails 3 in. back from the top of the chord. The gable-vent hole can be cut out at this point, and the vent tacked in for installing the siding. The vent is then removed to give us a hole for our lifting straps, and it is reinstalled later (photo p. 114). When there are no gable vents, we leave off the last couple of siding courses at the top of the truss and cut a hole for the straps. We precut the missing siding pieces and tack them to the gable end to be nailed on later from a ladder.

After the siding is nailed on, we snap a chalkline 2½ in. down from the top edge of the top chord and cut the siding off to that line. We nail on a piece of 1x3 above the cut siding as a spacer. The rake boards and band moldings are nailed over the 1x3, and the height of the rake boards is adjusted to match the thickness of the roof sheathing. We leave the tails of the rake boards long and cut them off after the gable truss is set in place.

If rake overhangs are called for, we take extra care to keep the top chord of the truss perfectly straight while we're sheathing the truss. We make our overhangs out of a 2x4 ladder with waste sheathing nailed to the topside of it to keep it stable. The overhang assembly is nailed to the sheathed gable truss, keeping the top of the assembly flush with the top of the truss. We install ½-in. AC plywood on the underside of the overhangs for our soffits and nail the rake boards and band moldings to the outside of the ladder, again adjusting for the thickness of the roof sheathing. We side the gable the same as before except that now the siding butts to the plywood soffit and a 1x2 frieze board is added.

We often have a chimney or chimney chase on the gable ends of our houses. Before siding the truss, we meet with the mason and determine the location of the chimney so that we can break back the siding and rake trim to his dimensions. After the gable truss is installed, we drop a plumb bob from the gable siding to the fireplace footing and snap a chalkline for the mason to go by. The final step in prepping the gable truss is starting nails in the bottom of the truss that will attach the bottom chord of the truss to the end-wall plate (bottom photo, p. 115).

Using a nailer saves a truss—After the trusses are ready, there are other preparations that make the raising go more smoothly. First we set up all of the staging front and rear. All of the walls that will receive the trusses are laid out, and strings are set up along the inside of the top plates for straightening the walls. We install adjustable diagonal braces to help straighten and tune the walls. We double-check the plans to see where the second-floor walls will be built in order to locate the best out-of-the-way places to drop materials (sidebar p. 119). We build a lot of two-story colonials with single-story attached garages. We save the cost of one truss by using a nailer in place of the truss that ordinarily would go against the wall of house (photo facing page). We slide one truss up before the crane arrives and use it as a template to mark the nailer location. After installing the nailer on the house wall, we set the truss in place and wait for the crane do the rest.

We also lay out ten 16-ft. pieces of 1x3 with 2-ft. centers for truss spacing and have another ten ready for diagonal bracing. We round up a bunch of clubs, 3-ft. to 4-ft. pieces of 2x4, to put under and between the piles of materials when the crane drops them off on the second floor.

Crew members have assigned tasks for the lift—As truss-raising day approaches, we watch the weather carefully. The best weather for raising trusses is calm and overcast. Wind is our worst enemy, and bright sun can make it difficult to see hand signals as well as trusses in midair.

We always shoot for an early start when the sun's angle is low and when the winds are usually calmest. Before the actual lifting begins, we round up all of the tools we'll need for the raising, including the truss spacers, a level and a straightedge for plumbing the first gable, and braces for holding and adjusting the first gable-end truss. We also get the tag line ready, fill our pouches with nails and recheck the walls to make sure they are straight. We also set up a ladder on the end of the house to make it easier to nail the bottom of the gable truss.

We always prefer to be a little overstaffed on raising day. Everyone has assigned duties, and the most important task is signaling the crane operator. We have used the same operator for six years, and we try to have the same crew member do the signaling for every raising (sidebar right). For safety's sake we always review the hand signals with the crane operator and the crew member who will be signaling. A thumb pointed in the wrong direction could get someone knocked off the staging in an instant. The signal person should never leave the crane operator's sight unless he signals "stop and hold."

For lifting materials, two crew members on the ground strap and launch loads, and two or three other crew members inside the house land and unstrap loads. The signal person usually sits in a second-floor window where he can watch materials coming in and see the crane operator clearly. After materials are secured on the second floor, we reposition the crew for the trusses.

The signal person is now stationed on the staging at the front of the house to set truss ends and to signal the crane operator. A second crew member is on the rear staging to orient the positioning marks on the trusses to the inside of the wall and to nail the trusses to the rear-wall plates. Two other crew members work in the middle of the trusses, tying them together at the peak when the crane drops them in. A crew member on the ground puts the trusses on the hook or strap and handles the tag line that steadies the trusses en route from the ground to the house. A sixth person is useful but not absolutely necessary to act as a gofer and to help when needed.

Lifting the first three trusses at once is safe and easy—Setting the first gable truss can be tricky. The best way we've found is sending that truss aloft with two other trusses, but with the gable truss in a separate strap (photo p. 116). If we sent up the gable truss by itself, we would have to depend on a brace with a steep angle to hold the gable until other trusses could be set. Our method lets the crane hold the gable truss steady until two other trusses are set and safer braces can be attached. Here's how it works.

After landing the first three trusses at the end of the house, the bottoms of the two regular trusses are kicked away from the gable about a foot so that they don't interfere with setting the gable truss. We adjust the gable truss to its positioning mark at the rear wall, then drive all of the nails we'd started earlier to fasten the bottom chord of the truss to the top plate of the end wall.

The strap holding the two regular trusses is then released, but the one holding the gable remains.

Getting the signals straight

Communication between the crane operator and crew is essential for safe and smooth lifting. A designated signal person has the responsibility of directing the lifting operation through hand signals. Here are the four basic hand signals we use with all of our crane operators.

Cable up/cable down— To move a load straight up or down, the crane operator must take up or let out the cable holding the load. To signal this procedure to the operator, point one finger either straight up or straight down and spin your hand in a circle.

Boom up/boom down— Raising the boom on the crane moves the load toward the crane, and lowering the boom sends the load away. The signal for this procedure is a thumb pointed either up or down with the hand moving in an up or down motion.

Boom right/boom left— Moving a load side to side is accomplished by pivoting the crane and boom. Pointing a finger to one side or the other in the direction you want the load to move is the proper signal for this procedure.

Stop and hold— To hold a load still momentarily, for example, while positioning blocks on the deck for materials to land on, a fist held stationary in the air stops the crane operation until the next signal is given.—*R. A. and M. G.*

We use a short length of 1x3 with the same 2-ft. layout as the wall plate to anchor the first two regular trusses to the gable. Then the bases are nailed to the wall plates on the rear wall.

Next, we brace the gable end with a couple of long diagonals running down to the subfloor. The extra crew member then slides two 2x braces through the webs of the regular trusses, and the top ends of the braces are nailed as high as possible to the gable-end truss. We make sure that the braces won't interfere with the bottom chord of the next truss we lift into place. We plumb the gable truss with a level and a straightedge, and when it's ready, the extra crew member fastens the braces to 2x blocks that have been nailed through the subfloor and into joists. Only after the braces have been nailed at both ends do we release the strap from the gable truss.

After the first trusses are set, we usually lift two trusses at a time using a hook that the crane operator had made. The rest of the trusses go up rather quickly. One of the two crew members working the center of the trusses holds the trusses while the other fastens the Truslock truss spacers (left photo, p. 117) (Truslock Inc., Route 1, Box 135, Calvert City, Ky. 42029; 800-334-9689). These spacers unfold, gripping the next truss and holding it at the proper spacing in one simple motion (see *FHB* #93, p. 98).

The two crew members working the middle of the trusses set the pace of installation and direct the others. The crew member at the rear of the house continues to align the positioning mark on each truss with the inside of the rear wall and nails his end of the truss securely with three 16d nails. The crew member at the front of the house just traps his end of the truss with a couple of nails (right photo, p. 117). After all of the trusses are up, we restraighten the walls and nail the front ends of the trusses with three 16d nails.

The extra crew member follows the action, strapping the ridge of the trusses with the 16-ft. 1x3s and tacking diagonal braces to secure the trusses until sheathing goes on. After the ridge is strapped, the truss spacers can be removed and used further down the roof. The top of the other gable end is tacked until we can set the distance between gables to the same measurement as the overall length of the house.

The garage trusses are a breeze by comparison. The first truss has been braced to the house wall. We just set the trusses the same as before and plumb the garage gable after the crane leaves.

Lifting the materials and setting the trusses for a 2,600-sq. ft. to 3,200-sq. ft. house with an attached garage generally takes us three to four hours. By the end of the day, we usually have the entire roof sheathed and our subfascias set with a crew of five or six. If for some reason we can't sheathe at least one side of the roof before we have to leave for the day, we put on plenty of diagonal bracing to secure the trusses in case the wind decides to kick up during the night. □

Rick Arnold and Mike Guertin are partners in Midcor Construction Inc., a building company, as well as U. S. Building Concepts Inc., a construction consulting business. Photos by Roe A. Osborn except where noted.

Let a crane do the heavy lifting

No matter what your physical condition, reducing the amount of bull work on the job is always welcome. A crane is one great way of saving muscles as well as man-hours and money (our crane operator usually charges around $65 per hour). Before the trusses are raised, the crane lifts the roof sheathing, second-floor interior-wall studs, furring strips and roof shingles to the second floor.

Materials are prepped for lifting ahead of time—With a rough count of all of the second-floor interior studs we'll need, we crown, cut and stack all of the framing stock on 3-ft. to 4-ft. scraps of 2x4 (top photo). The roof shingles are usually delivered the day before the truss raising, along with several extra pallets to stack them on. We generally stack 21 to 24 bundles per pallet and use a long hook that the crane operator made to feed the lifting strap through the pallet. We use 3-ft. 2x4s as strap spreaders to keep the lifting straps from crunching the top bundles of shingles (center photo). Roofing subcontractors love having the shingles on the second floor. Passing the bundles out a window, onto the staging and up to the roof is much easier and safer than hauling them up two stories on a ladder.

The roof sheathing is sent out at the same time as the shingles. Our lumberyard will band sheets of plywood in any quantity that we ask. Bundles of 15 to 20 sheets for ⅝-in. plywood or bundles of 20 to 25 sheets for ½-in. plywood seem to work best for us (bottom photo). When the sheathing is slid off the delivery truck, we ask the driver to put heavy-duty nylon load straps around the entire stack to keep the steel bands from snapping. Occasionally, we get lucky and are able to coordinate delivery of the sheathing and shingles with the actual lift. In those rare instances the crane plucks the sheathing and shingles right off the delivery truck and lifts them up to the second floor.

Placement of the materials on the second-floor deck takes a bit of planning—Roof sheathing and the furring will be used up before the interior walls are built so that material can go almost anywhere. The studs and the roof shingles, however, need to be positioned so that they won't interfere with framing the interior walls. We try to land the stacks near windows for easy handling, and because of the concentrated weight, we keep them away from the middle of the floor joists to avoid overloading them.

When rigging the bundles and stacks with the lifting straps, we try to keep the straps as far apart on the load as possible. The crane operator takes the strain slowly so that we can make sure the load is going to stay flat and even while it's in the air. It's not uncommon to have a load put back down on the ground to get the straps positioned just right. This minor hassle is infinitely preferable to a load coming apart or straps shifting and sliding while a load is in midair.—*R. A. and M. G.*

Interior wall framing on the rise. Lumber is stacked ahead of time for an easy crane hoist; crew members land the framing lumber inside the house where it will be out of the way until it's needed.

Roof shingles are dropped near a window. The crew member on the left signals the crane operator while the other crew member guides the pallet of shingles to its temporary home.

Prepackaged loads of sheathing are easier to handle. The lumber company bundles the sheathing in stacks of 15 to 20 sheets for easy strapping and lifting.

The hip goes up in one piece. Like a giant box kite on a string, an entire hip section is lifted to its home atop a two-story house.

Building Hip and Valley Roofs With Trusses

Keep the numbered trusses in order, and entire hip sections can be assembled on the ground and lifted as single units

by Rick Arnold and Mike Guertin

A few years and many roofs ago, a builder approached us about framing a two-story colonial-style house. He said that he wanted to try a truss system for the hip roof. We had used trusses for a lot of gable roofs, but we had never seen a hip done that way. When the trusses for the job were delivered, we just stood back and scratched our heads. It looked as if bunches of unrelated pieces had been strapped together in no particular order. The engineering plan looked like a map of some unfamiliar suburb.

Too smug to admit that we needed help, we muddled our way through, lifting each weird-looking truss up to the roof by hand and then moving each piece three or four times until we found the right spot to nail it. That roof is still in good shape after two hurricanes, and whenever we drive by the house, we chuckle at how much time and effort it took to put that roof together.

Our methods for assembling hip-and-valley truss systems have evolved a great deal since that first puzzled attempt. The biggest advance

in our technique came when we described the process to our crane operator. He suggested assembling some of the trusses on the ground and lifting whole hip sections onto the house in one shot (photo above). It worked like a charm. Now we even sheathe the assemblies before they go up.

Building hip systems on the ground is quicker and safer—Before anything is assembled, we prep the hip-and-valley trusses much as we do standard trusses (see pp. 114-119). We line them up in a stack on the ground and mark layout lines for the sheathing and strapping and for alignment on the walls. If necessary, we also restack the trusses that will be lifted individually by the crane to be sure they're in the proper order. We lay out the wall plates according to the truss plan (sidebar p. 123) and write the number and designation for each truss at its layout point.

When framing hip roofs with trusses, we most often use a step-down hip truss system (top drawing). Trusses in this system have the same span as common trusses, but they're flat on top (for more information on types of trusses, see the drawing, p. 105). The flat parts of the hip trusses become progressively wider and lower as the trusses step away from the last common truss to begin forming the hip. The lowest and widest hip truss, the hip-girder truss, supports a series of monotrusses, called jack trusses, that complete the roof. The hip-girder truss usually has a heavier bottom chord than the other hip trusses to accommodate the extra weight of the jack trusses and the metal hangers that hold the jacks. Generally, two girder trusses are nailed together and work in tandem for each hip system.

After the wall plates are laid out, we lay out a hip-girder truss while it is still lying flat on top of the pile. We begin our layout by locating the exact vertical center of the truss, top chord to bottom (bottom drawing). First we locate the middle of the top flat chord of the truss. Then from the ends of the flat chord, we measure equal distances diagonally to the bottom chord. Halfway between our diagonal marks is the midpoint of the bottom chord. The line between the midpoints is the vertical centerline.

From this centerline we can locate the position of the outermost jack trusses on both the top and bottom chords as indicated on our truss plans. A jack truss is a monotruss with a single top chord. There are three different types of jack trusses in a hip system: face jacks that are attached to the face of the hip-girder truss and that run perpendicular to it; king jacks that run diagonally from the girder truss and form the outside corners of the roof; and side jacks that are attached to both sides of king jack trusses.

The layout for the rest of the face jacks is taken from the wall-plate alignment mark on the girder truss. However, the center jack truss is always at the exact center of the girder, regardless of the spacing. We tack metal hangers for the face jack trusses onto the bottom chord of the girder truss with just a couple of nails in each hanger. They will be nailed in permanently with spikes after the second girder truss is mated to the first.

Jack trusses are nailed to a pair of hip-girder trusses—Next we move the prepared hip-girder truss to a relatively flat area of the job

Drawings: Chuck Lockhart

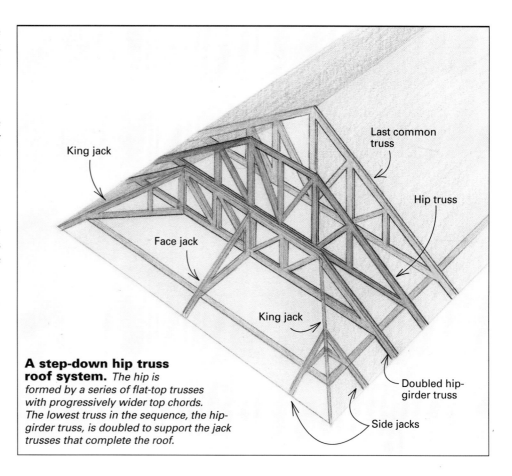

King jack

Last common truss

Hip truss

Face jack

King jack

Doubled hip-girder truss

Side jacks

A step-down hip truss roof system. *The hip is formed by a series of flat-top trusses with progressively wider top chords. The lowest truss in the sequence, the hip-girder truss, is doubled to support the jack trusses that complete the roof.*

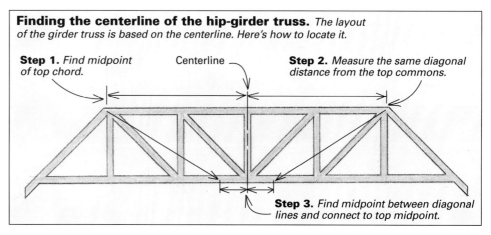

Finding the centerline of the hip-girder truss. *The layout of the girder truss is based on the centerline. Here's how to locate it.*

Step 1. *Find midpoint of top chord.*

Centerline

Step 2. *Measure the same diagonal distance from the top commons.*

Step 3. *Find midpoint between diagonal lines and connect to top midpoint.*

A jack truss holds the girder truss upright. The center face jack truss is tacked to the girder truss to keep it vertical while the rest of the face jacks are installed.

Straightening the girder trusses. A temporary brace keeps the girder trusses straight while they are being nailed together.

A furring-strip spacer keeps the tails in place. Before the system can be squared, the tails of the face jack trusses are spaced according to the layout and held in position with a piece of 1x3.

Diagonal measurements square the system. Measurements are taken between the two outermost jack trusses, and the tails of the trusses are moved in unison until the measurements are equal.

site and prop it upright on blocks. The center face jack truss is set into its hanger and tacked to the girder truss at the top to hold them both upright (photo left, p. 121). The tails of all of the jack trusses need to be supported, so we make a continuous block out of long lengths of 2x material. The blocking for the jacks is raised until the hip-girder truss is sitting fairly plumb, and the rest of the face jacks are then slipped into their hangers and tacked at the top. When they're all in place, we nail them off through the chords and webs of the girder truss.

Now we tack the second hip-girder truss to the first with just a few nails so that the girder trusses can be straightened before they are joined permanently. We run a stringline on the top and bottom chords of the girder trusses to get them straight. If need be, we temporarily brace the bottom chord against the ground to keep it straight (photo right, p. 121). The tails of the face jacks are kept at the correct spacing with furring strips, marked to match the jack-truss layout and tacked on top of the bottom chords (photo left). When the girder trusses are straight and the face jack trusses are spaced properly, we nail the two girder trusses together through all of the chords and webs, and we nail off the hangers for the face jacks.

We are now ready to square the assembly. First we recheck our strings on the girder trusses and then measure diagonally between the top chords of the two outermost face jack trusses (photo right). The tail ends of the face jacks are moved in unison until our measurements are equal and the face jacks are square with the girder truss. We check our strings one last time and nail a furring strip diagonally onto the underside of the top chords of the face jacks and on top of their bottom chords to keep the whole system square and uniform.

A 2x4 subfascia is now nailed to the tails of the face jack trusses and extended far enough to catch the tails of the king jack trusses when they're installed. We will straighten the subfascias after the hip systems are installed on the house walls. If the hip roof is going on a single-story house, we ordinarily stop here. Because the staging is simpler for a single-story house and because materials can be passed to the roof directly from the ground, it's quicker for us to complete the assembly in place.

Special hangers hold the jack trusses for the hip corners—For multistory houses we finish building the hip section on the ground. The next step is tacking the king jack trusses in place (photo top left, facing page). The king jack truss is built with the top chord at the same pitch as a hip rafter and functions in much the same way. The king jack trusses are installed between the hip-girder truss and the last face jack truss on both ends of the assembly. They fit into a specially designed hanger, provided by the truss manufacturer, that eliminates the need for the

45° angles normally cut on the top end of a hip rafter. We position the king jacks at exactly the same distance from the girder truss and the adjacent face jack truss, and we temporarily hold them in place with a furring-strip brace.

The tails of the hip trusses are usually left long by the manufacturer and cut to length on site. We run a string along the tails of the face jacks to determine where the tails of the king jacks need to be cut. We usually make this cut with a reciprocating saw because the subfascia tends to get in the way of a circular saw.

Next we cut the return angle on the tail of the first king jack. We find this cut by measuring along the bottom chord of the hip-girder truss from the first face jack truss to the end of the overhang. Then we measure that distance from the tail of the face jack to the tail of the king jack and make our cut there. The subfascia is cut to the same length and then nailed to the king-jack tail. On the other end of the assembly, we cut the king-jack tail and subfascia so that the length of the subfascia is the same as the overall length of the girder truss. The returning subfascias can now be nailed on. If possible, we extend the returning subfascias back beyond the girder trusses to tie into the other step-down hip trusses when they are installed.

The side jack trusses are attached directly to the king jacks (photo top right, facing page). A side jack truss is a simple monotruss with just a top and a bottom chord joined together with

The king jack truss forms the corner. The top chord of the king jack is cut to the pitch of a hip rafter and functions similarly. It is held in place with a special hanger, and the tail is positioned equidistant from the girder truss and the outermost face jack.

Side jacks fill in the framing beside the king jack. Side jack trusses consisting of just a top and bottom chord are nailed to the king jack and the subfascia.

Truss fabricators engineer the roof for you

Truss systems can be designed for almost any complicated roof design (photo right). Although the cost for the truss package may be more than the cost for conventional framing materials, the labor savings are phenomenal. As an added benefit, the interior bearing walls necessary for conventionally framed roofs can be eliminated, which allows greater design flexibility. Some of the truss packages we have ordered even include complicated details such as vaulted ceilings and roofs with hips and valleys of different pitches, roofs that would have been a real challenge to frame conventionally.

We've found that the best way to explore the possibilities is to go over the house plans with the engineer for the truss fabricator. At that meeting we often make arrangements for minor structural changes in the house to accommodate the roof trusses. Sometimes by moving a couple of bearing points or inserting a carrying beam, we can change a roof from conventional framing to a truss system. The benefits we gain by using trusses have always outweighed any changes we need to make.

After the meeting our fabricator usually gets back to us within a few days with a rendering of the truss plan and a price. We are occasionally astonished when a truss system ends up costing less than the lumber for a conventionally framed roof because trusses are made of less expensive, smaller dimension lumber.

We study the truss plan before the trusses arrive. If we're having trouble understanding a plan view of the system, the truss fabricator will provide an isometric drawing, usually at no extra cost. An isometric drawing shows the roof in 3-D and helps clarify the more difficult details.—*R. A. and M. G.*

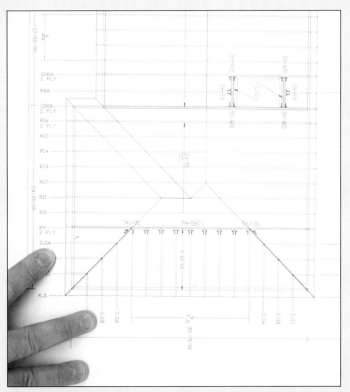

The map of a complicated roof system. Engineered-truss plans like this one, which is for the roof pictured on p. 81, are provided by the truss manufacturer. The plan identifies each type of truss and its exact location. This roof features a dozen different types of trusses.

A valley kit is nailed directly to the sheathing. A series of progressively smaller trusses, called a valley kit, forms the valleys between two intersecting roofs.

truss plates at the splice point. We locate the attaching points for side jacks by pulling 24-in. centers off adjacent girders and face jack trusses. The subfascias are laid out the same way.

Usually the manufacturer cuts the side jack trusses to the proper length but without the 45° angles on the ends of the chords. We cut these angles on site. For each hip system, there are four of each size side jack truss, two with right-hand 45° cuts and two with left. Having one person organize and cut the side jacks minimizes the chances of cutting them wrong. Once the side jacks are cut with the proper angle, they can be positioned and nailed to the king jack truss and the subfascia.

We sheathe the trusses by snapping lines across all of the jack trusses and filling in as many sheets as we can. We don't sheathe return facets of the hips until the system is in place on the house so that our sheathing will tie back into the other trusses. Lower sheets are tacked in place temporarily so that we can lift them out of the way when we nail trusses to top plates.

Assembled hip sections are lifted level—
When we lift the hip sections, we run heavy-duty straps through the top corners of the hip-girder trusses where the king jack trusses are attached for the strongest lifting points (photo p. 120). We attach an adjustable strap around the tails of the three middle face jack trusses to balance the load. The crane operator lifts the assembly just a little so that we can adjust the middle strap and get the assembly as level as possible before it's lifted into place. The more level the assembly is, the easier it is to position on the walls.

When the assembled hip system is airborne, a crew member stationed on the ground keeps it steady with a tag line until it is within reach of the crew on the staging. First we land the assembly at the layout marks on the plates for the hip-girder trusses. Then we have the crane tug the whole assembly toward the front or back until the lines on the bottom chord of the girders align with the inside edge of the walls. When the system is properly positioned, we release the straps and nail the trusses to the top plates of the end wall and along one side. The process is repeated for the opposite hip assembly.

The step-down hip trusses are now lifted into position one at time. We tack on short pieces of furring just below each of the flat sections to connect one truss to the next at the correct spacing. The process is repeated until we set the first full-height common truss. Then we pause the crane and tack a long piece of furring onto the flat tops of the trusses, measuring and spacing them properly as we go.

The common trusses are sent up two at a time and locked in place temporarily with Truslock truss spacers (Truslock Inc., 2176 Old Calvert Road, Calvert City, Ky. 42029; 800-334-9689; see *FHB* #93, p. 98). The layout for the last common truss is usually irregular to accommodate the step-down hip or valley system. We space that truss with a piece of furring marked to reflect the difference in the layout. The step-down hip trusses for the other end of the house can now be sent up one at a time and braced with furring as we did with the first end.

Building valleys out of trusses is a snap—
There are two basic ways of framing valleys with trusses. The one we encounter most frequently, and the easiest to frame, is the intersection of two simple roofs. We begin by setting all of the

trusses for the main roof. If there are no interior bearing walls to support the trusses of the main roof where the two roofs meet, we hang the ends of the unsupported trusses on hangers nailed to a girder truss, which is part of the other intersecting roof.

Once the main roof has been sheathed, we set the trusses for the intersecting roof as far as the main roof. The valleys are created with a valley kit, which is a set of progressively smaller common trusses nailed directly onto the sheathing (photo left). We usually rip the pitch angle of the intersecting roof onto the bottom chord of each truss in the valley kit. Although cutting trusses is not a practice that is generally accepted, our truss manufacturer has assured us that ripping valley-kit trusses for this type of application is permissible.

If a valley is so close to a hip that a girder truss can't be used to hold the main roof trusses, a second method using special step-down valley trusses may be the answer (photos and drawing, facing page). These valley trusses are similar to common roof trusses except that the top chord is interrupted by a flat extension. The length of the flat extension is the same for every step-down valley truss and reflects the distance between the hip and valley lines. However, the height of the flat extension increases with each successive truss, creating both the valley and the hip as they go. Again, because each truss in the series is different, we take extra care to stack the trusses in the order that they will be lifted by the crane. Following the truss plan to the letter helps a great deal.

All of the hip and valley lines are reinforced with blocking—After all of the trusses are set and the crane leaves, we complete the truss installation with a few extra details. First, we beef up the hips and valleys with 2x blocks cut with compound angles. These reinforcing blocks are not specified on the engineer's plan, but nailed between the step-down trusses at the hip and valley lines, they're helpful as spacers for the trusses and as nailers for the sheathing.

As with a simple truss roof, we nail the trusses only along one side of the house during the raising. After the crane leaves, we restraighten the walls in case they've been knocked out of line while the trusses were set. When the walls are straight, we nail off the other end of the trusses. The rest of the subfascias can now be installed and shimmed straight. After our lunch break, we finish sheathing the roof. □

Rick Arnold and Mike Guertin are partners in Midcor Construction, a building company, as well as U. S. Building Concepts, a construction consulting business in East Greenwich, Rhode Island. Photos by Roe A. Osborn except where noted.

Photo this page: Mike Guertin

Hips and valleys next to each other require special trusses. *When a hip and a valley are too close together to be framed with conventional trusses, as in the drawing below, special step-down valley trusses have to be used.*

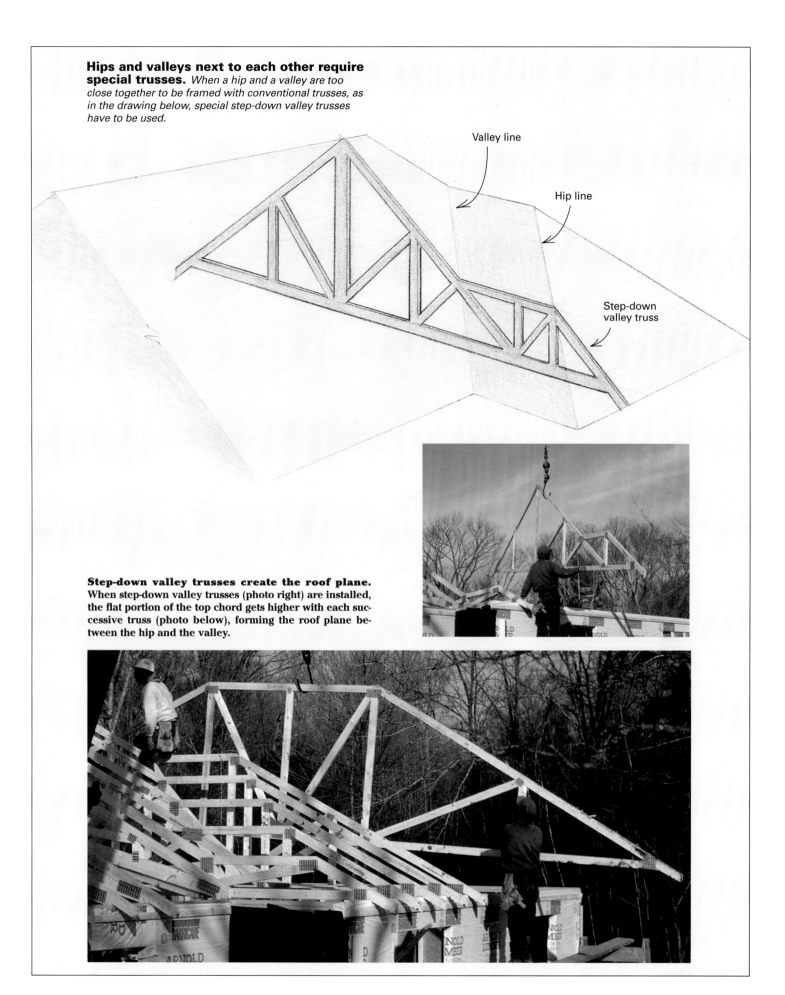

Valley line

Hip line

Step-down valley truss

Step-down valley trusses create the roof plane. When step-down valley trusses (photo right) are installed, the flat portion of the top chord gets higher with each successive truss (photo below), forming the roof plane between the hip and the valley.

Strengthening Plate-to-Rafter Connections

It may be time to abandon the time-honored toenail

Failure of a toenailed connection.

by Stanley H. Niu

Overloaded rafter-tie connection.

Overloaded lag-screw connection.

Late September 1989: Hurricane Hugo clobbers South Carolina. April 26, 1991: Tornadoes knife through Butter County, Kansas. In these and many other instances of extreme weather, wood-frame houses are among the most heavily damaged structures. The mode of failure is predictable: The roof blows off, leaving bare walls to weather the storm.

Many people consider the damage caused by hurricanes and tornadoes to be an act of fate and assume that nothing can be done to prevent the destruction. Perhaps this is true, but I'm convinced that the damage can certainly be reduced, and with minimal expense. If the roof

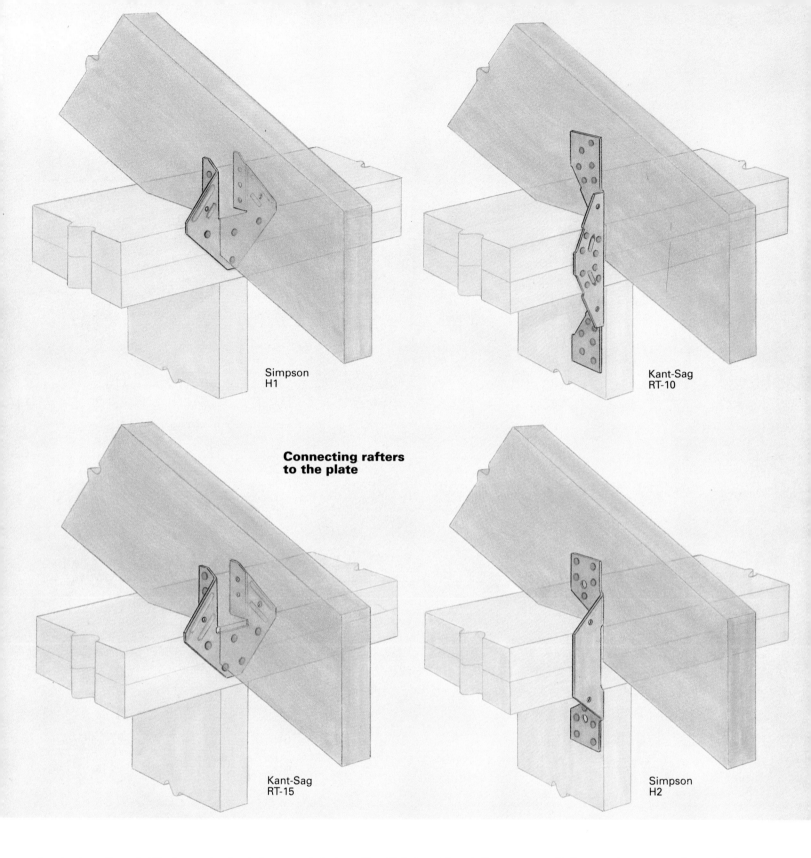

Connecting rafters to the plate

Simpson
H1

Kant-Sag
RT-10

Kant-Sag
RT-15

Simpson
H2

stays in place, the rest of the house stands a better chance of resisting the storm. The key is to improve the strength of the connection between the top plates and the rafters. Research I recently undertook with colleagues Laurence Canfield and Henry Liu shows just how much this connection can be improved.

Making a better connection—Although wind pressures on buildings have been studied extensively, only a few studies have examined the strength of the rafter/plate connection. What is known, however, is that a connection made with metal rafter ties is considerably stronger than one made by toenailing. Unfortunately, not all manufacturers publish information regarding the maximum recommended uplift loads their ties can resist. Without those figures, it's tough to pick the appropriate tie. So in our laboratory, we tested a selection of ties in various shapes from several manufacturers (chart, p. 129) to establish the ultimate strength of each tie. We also investigated the uplift resistance of toenailed connections, as well as two different sizes of lag-screw connections (the lags were run through the rafter and 3 in. into the plates; a washer was included).

First, I'll give a couple of notes about our testing procedures. Nails used for the three different toe-nailed connections we tested included 8d common nails, 16d box nails and ring-shanked, 16d common pole barn nails. The 16d nails often split the rafter during nailing, so we predrilled the rafters with a $\frac{5}{32}$-in.-dia. hole. It is unlikely, however, that carpenters would drill pilot holes in the field. Each toenailed connection used three nails: two on one side and one centered on the other side. The lumber we used for all tests was construction-grade stock obtained from a job site, and we inspected it to ensure that no flaws or cracks would bias the test results. After the appropriate rafter/plate connection was made, samples of the assembly were placed in a

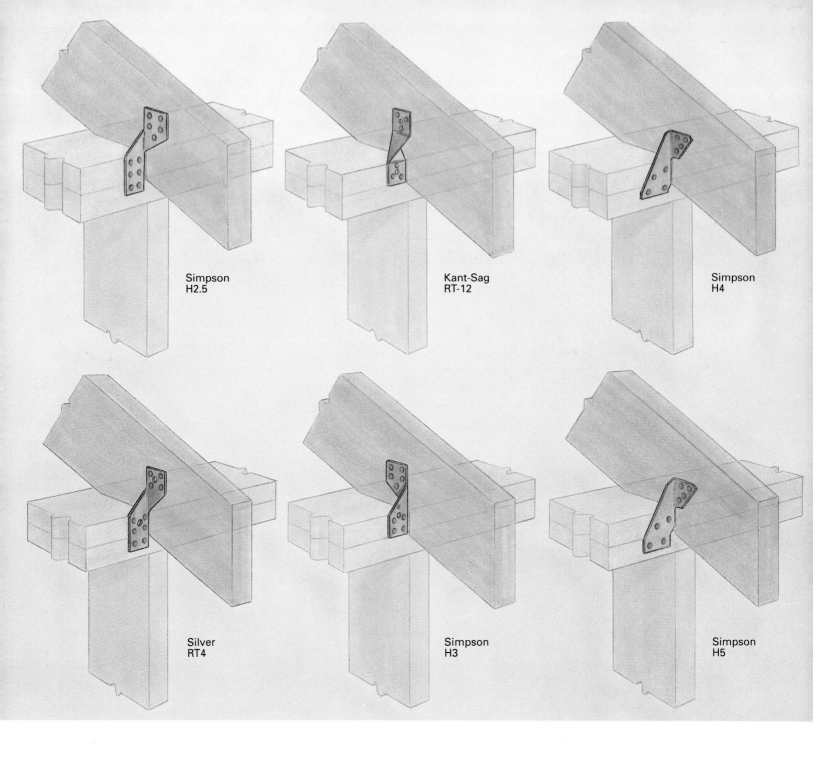

Simpson
H2.5

Kant-Sag
RT-12

Simpson
H4

Silver
RT4

Simpson
H3

Simpson
H5

hydraulic test apparatus that pulled the rafter away from the plate. We tested at least 15 connections, pulling until the connection failed.

The results of these tests are shown in the chart on the facing page. Ties fell into three groups ranked according to their average load capacity: below 650 lb.; 900 lb. to 1,300 lb.; and above 2,700 lb. (the last group represents the high-performance end of the spectrum, with a load capacity that is double or triple most of the midrange connections). As it turns out, the weakest sample tested was the 8d toenail connection, with an average load capacity of only 208 lb. In contrast, the lowest-capacity rafter tie tested had an average load capacity of 497 lb. When the toenailed connection failed, the nails pulled out of the top plate (top photo, p. 126). In some cases, when the connection failed, the bottom of the rafter split first. When a metal tie fails, it usually tears in half, but the nails stay put (bottom left

photo, p. 126). The lag-screw connections failed when the lag pulled free of the top plate (bottom right photo, p. 126). Unfortunately, toenailed rafters are probably the most common rafter/ plate connection found in wood-frame houses. In fact, this connection is in compliance with the Uniform Building Code (UBC) and the Building Officials and Code Administrators (BOCA).

Applying the research—Putting our results to the test on a hypothetical house shows how important it can be to use the right connection. Consider a house located near Kansas City, Missouri, with a 30-ft. by 60-ft. floor plan and a hip roof. The rafters are located 16 in. o. c., which calls for a total of 86 rafter connections, and the roof has a 3-in-12 pitch with no overhang. The house is located on open terrain surrounded by scattered obstructions having heights of 30 ft. or less. A map of wind speeds shows that the Kansas

City area has a basic wind speed of 75 mph (basic wind speed, an engineering term, is the fastest wind speed measured at 33 ft. above the ground with a 2% annual probability of occurrence).

For an 1,800-sq.-ft. roof, the total wind lift on our hypothetical house equals 31,824 lb. Dividing this by 86 connections yields a 370-lb. uplift load per connection. Based on our test results, any of the ties tested would be adequate for this region. However, a connection made with three 8d common nails has a load capacity of only 208 lb. In some weather conditions, this connection would be inadequate.

Now consider the same house located on oceanfront property in South Carolina. The basic wind speed there is 100 mph, so the load per connector would be 957 lb. on the same roof. Metal connectors from the middle group (load capacity from 900 lb. to 1,300 lb.) would be adequate, though some barely so. However, wind speeds

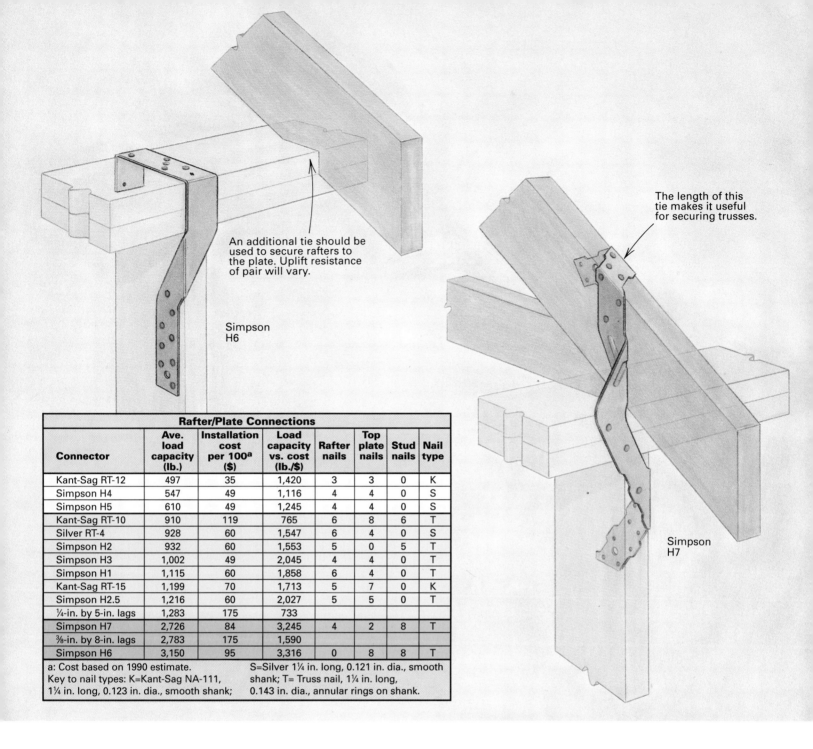

An additional tie should be used to secure rafters to the plate. Uplift resistance of pair will vary.

Simpson H6

The length of this tie makes it useful for securing trusses.

Simpson H7

Rafter/Plate Connections							
Connector	Ave. load capacity (lb.)	Installation cost per 100[a] ($)	Load capacity vs. cost (lb./$)	Rafter nails	Top plate nails	Stud nails	Nail type
Kant-Sag RT-12	497	35	1,420	3	3	0	K
Simpson H4	547	49	1,116	4	4	0	S
Simpson H5	610	49	1,245	4	4	0	S
Kant-Sag RT-10	910	119	765	6	8	6	T
Silver RT-4	928	60	1,547	6	4	0	S
Simpson H2	932	60	1,553	5	0	5	T
Simpson H3	1,002	49	2,045	4	4	0	T
Simpson H1	1,115	60	1,858	6	4	0	T
Kant-Sag RT-15	1,199	70	1,713	5	7	0	K
Simpson H2.5	1,216	60	2,027	5	5	0	T
¼-in. by 5-in. lags	1,283	175	733				
Simpson H7	2,726	84	3,245	4	2	8	T
⅜-in. by 8-in. lags	2,783	175	1,590				
Simpson H6	3,150	95	3,316	0	8	8	T

a: Cost based on 1990 estimate.
Key to nail types: K=Kant-Sag NA-111, 1¼ in. long, 0.123 in. dia., smooth shank; S=Silver 1¼ in. long, 0.121 in. dia., smooth shank; T= Truss nail, 1¼ in. long, 0.143 in. dia., annular rings on shank.

of 125 mph were reported during Hurricane Hugo, and lifting loads during the storm would have been 1,496 lb. per connector. Only the top two connections would have been adequate: the H7 tie and the ⅜-in. by 8-in. lag screw. Of course, the rest of the structure would require sufficient strength to prevent it from being blown off the foundation. But either of these connections would have improved the chances of keeping the roof in place.

The cost of safety—Our research was done in a laboratory, so it was easy to see which connection would be the best. But on the job site, "best" often competes with "cost-effective" for the right to determine what gets built. That's why we calculated the installed cost of each connection. In determining the costs, we assumed that a carpenter would take 10 seconds to install each nail and would earn an average wage of $21 per hour.

The average house would probably require from 80 to 120 connectors. As you can see from the chart above, the additional cost incurred by using rafter ties is negligible compared to the total cost of the house.

Manufacturer's guidelines suggest that ties be installed with at least four nails each to prevent the tie from rotating. However, more nails ensure a better connection.

Improving the improvements—Though the rafter ties performed well as a group, we identified some modifications that could improve their uplift strength. The H4 and H5 rafter ties could be made from 18-ga. sheet metal instead of the 20-ga. sheet metal currently used, and the nail holes could be slightly larger to accommodate truss nails. The H2 rafter tie has a hole on its face between the rafter and the top plate. In our tests, the tie failed by tearing in half, with the tear starting on the inside edge of the tie and progressing to this hole. Elimination of the hole might improve the strength of the tie. The RT-10 rafter tie, which is similar to the H2, also failed by tearing in half between the rafter and the top plate. This tie would be improved if it were wider (more like the proportions of the H2). Generally, the 18-ga. sheet metal used for most of the ties seems a good compromise between strength and low manufacturing cost. □

Stanley H. Niu is an associate professor in the Department of Civil Engineering at the University of Missouri. Professor Henry Liu supervised the research. Laurence Canfield, plant engineer at the Wire Rope Corp. of America (St. Joseph, Mo.), conducted the experiments. Photos by the author. For further information on the methodology of these tests, see the Forest Products Journal, *July/August 1991 (2801 Marshall Ct., Madison, Wisc. 53705).*

Cornice Construction

Building the return is the tough part

by Bob Syvanen

All gable-roofed houses need to have a cornice of some sort. Functionally, the cornice fills the voids between roof and sidewall. It extends from the shingles to the frieze that covers the top edge of the siding. There are two basic kinds of cornices: one includes a gutter (drawing **1A**, below); the other is a simple cornice molding (**1B**). Traditionally, each calls for a different method of roof framing. For the gutter, rafter tails are combination-cut (plumb and level) and bear on the top plates. For a cornice without the gutter, a double plate is nailed across the tops of the joists to support the rafters. There are a vast

number of possibilities in cornice construction and detailing (photos, facing page), but they are all variations on the basic, step-by-step sequences shown and discussed here.

The cornice return at the juncture of eave and gable is the most noticeable and most intricate part of this architectural detail. Sometimes called a boxed return, it adds a delicate touch to the large, repetitive detail of clapboard or shingle siding.

Building the cornice itself is simple; building its return is a bit more challenging. Let's start with the eave corner, assuming that you've already framed and sheathed it. The

rafter ends and ceiling joists should still be exposed, and the first step is to nail up the fascia board. For the guttered cornice (**2A**), the fascia will go against the rafter tails; otherwise, nail it to the ends of the ceiling joists (**2B**). For the fascia, I use #2 pine boards ripped to width on a table saw. The top edge of the fascia usually needs to be beveled for a tight fit.

At the corner of the house, the fascia meets the ear board—the broad, flat backing board at the gable base that holds the cornice return. The joint between the fascia and the ear board has to be mitered, and unless the fit is

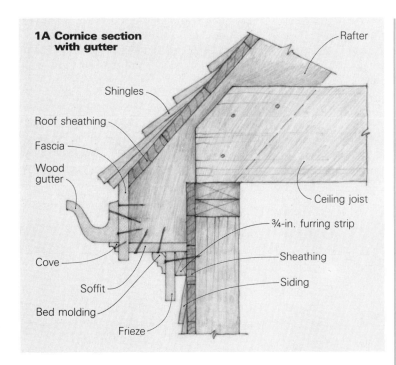

1A Cornice section with gutter

Rafter
Shingles
Roof sheathing
Fascia
Wood gutter
Cove
Soffit
Bed molding
Frieze
Ceiling joist
¾-in. furring strip
Sheathing
Siding

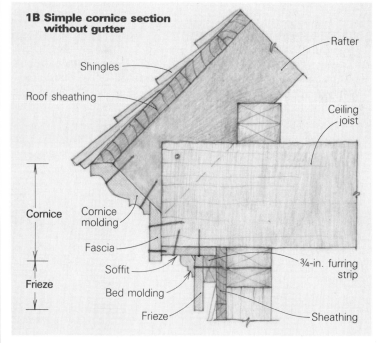

1B Simple cornice section without gutter

Rafter
Shingles
Roof sheathing
Ceiling joist
Cornice
Cornice molding
Fascia
Soffit
Bed molding
Frieze
Frieze
¾-in. furring strip
Sheathing

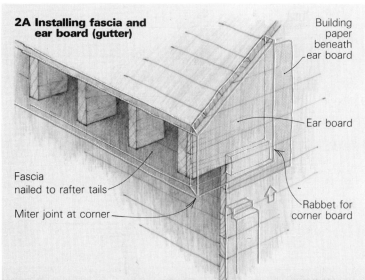

2A Installing fascia and ear board (gutter)

Building paper beneath ear board
Ear board
Fascia nailed to rafter tails
Miter joint at corner
Rabbet for corner board

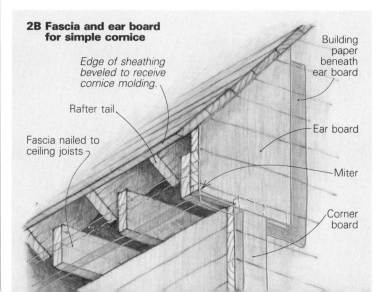

2B Fascia and ear board for simple cornice

Building paper beneath ear board
Edge of sheathing beveled to receive cornice molding.
Rafter tail
Fascia nailed to ceiling joists
Ear board
Miter
Corner board

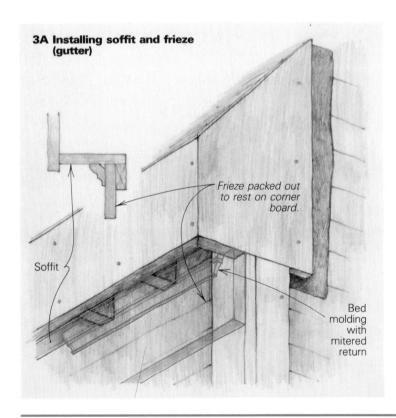

3A Installing soffit and frieze (gutter)

Frieze packed out to rest on corner board.

Soffit

Bed molding with mitered return

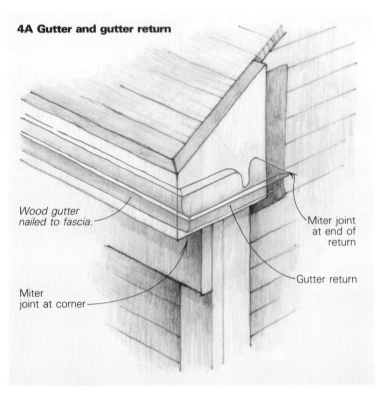

4A Gutter and gutter return

Wood gutter nailed to fascia.

Miter joint at end of return

Gutter return

Miter joint at corner

tight, it's a good idea to use a weatherproof adhesive caulk. Phenoseal (made by Gloucester Co., Box 428, Franklin, Mass. 02038) and other marine adhesives work well in this type of joint. I also staple a piece of building paper behind the ear board.

If you want the ear board to be flush with the corner board, which is installed before the siding is nailed up, the ear board needs to be rabbeted at its bottom edge so that the siding and a tongue at the top of the corner board can be tucked up underneath it. The ear board should also cover the gable edge of the roof sheathing if you're not planning to use

rake molding, which is shown in the inset drawing **6B**.

Now install the soffit, frieze and bed molding, in that order. I cut the soffit and frieze boards from pine or fir, and fur out the frieze a full ¾ in. so that the siding will fit underneath it. Cut the frieze to extend over the corner board, as shown in **3A** and **3B**. You can use something fancier than bed molding to cover the joint between soffit and frieze, or simply leave this juncture plain.

The return—The next piece to go on is the cornice (crown) molding (**4B**), or the wood

gutter that replaces it on a guttered cornice (**4A**). Here again, the corner is mitered to make the return. In fitting the return, I like to tack a temporary guide strip to the ear board, on a level line where the bottom of the cornice molding or gutter will fit. This makes test-fitting, trimming, and retesting the miter a bit faster.

A good tight joint here is important. Corners are seldom perfectly square, and you can either adjust the cut in the miter box or just keep fitting and trimming with a chisel or block plane. The small triangular piece at the end of the return is delicate, and I usually cut

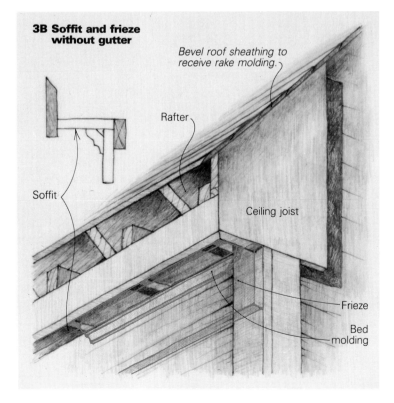

3B Soffit and frieze without gutter

Bevel roof sheathing to receive rake molding.

Rafter

Ceiling joist

Soffit

Frieze

Bed molding

4B Cornice and cornice return

Miter joints at corner and end

Cornice or crown molding

5A Water table over gutter

1 ¾ in.   ¾ in.

The water table, of 8/4 stock, is made ¼ in. longer and wider than the return to create a drip edge. It slopes away from the house.

6A Flashing and rake board

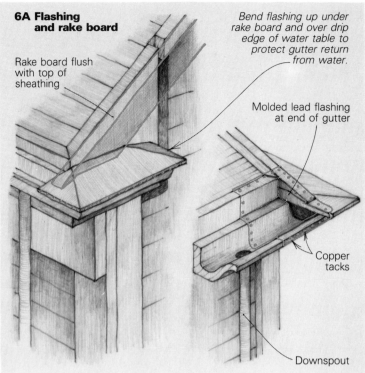

Rake board flush with top of sheathing

Bend flashing up under rake board and over drip edge of water table to protect gutter return from water.

Molded lead flashing at end of gutter

Copper tacks

Downspout

---

and fit it on the ground after the corner miter is done.

What you have now is the mitered return with an opening on top. This opening must be covered with a water table (**5A**, **5B**), which is a piece of 8/4 pine, beveled away from the side of the house to shed water. You can cut a single bevel, or bevel the board downward on all open sides for a fancier look. Size the board a little longer and wider than the return to create a drip edge. The water table for the simple cornice return has to be trimmed to fit beneath the roof sheathing (inset, **5B**).

Once you've nailed the water table in place,

flash it before installing the rake board along the corner between the roof and the gable wall. The flashing has to extend up beneath the rake board, so if you're working on an old house, pry the rake board up, tuck the flashing underneath it, and then renail the rake. If you are going to use rake molding (**6B**), nail the rake board to the gable end, furring it out away from the sheathing at least ¾ in. so that the siding can be fitted underneath. To provide good bearing for the rake molding, bevel the gable-end sheathing. If you don't use a rake molding, leave the sheathing square-cut so that the rake board and its furring strip

(**6A**) can be nailed to it. In either system, the rake board and molding die onto the water table, and must be scribed accordingly.

If you're building a guttered return, there's one final, important step: sealing off the corner. Water that gets into the return side of the gutter will soon rot the wood. Lead is the best flashing material because it can be formed to the gutter contours more readily than copper or zinc. Use the end of your hammer handle to form the lead flashing so that it covers the corner and fits over the water table, rake board, and roof sheathing. Then nail it down with copper tacks in a bed of caulking. □

5B Water table over simple cornice return

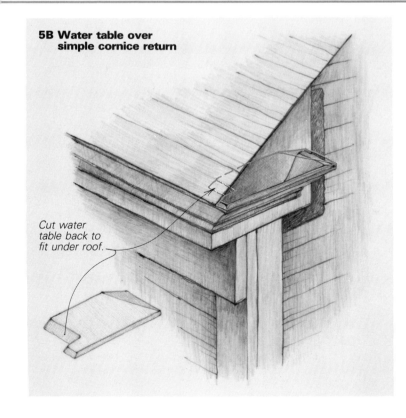

Cut water table back to fit under roof.

6B Flashing, rake board and rake molding

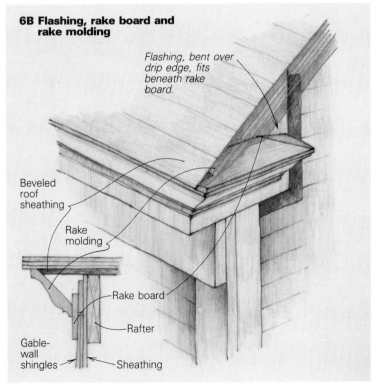

Flashing, bent over drip edge, fits beneath rake board.

Beveled roof sheathing

Rake molding

Rake board

Rafter

Gable-wall shingles

Sheathing

Illustrations: Frances Ashforth

Extending the rake. For a clean look, the top edge of the rake's crown molding extends down past the edge of the roof, where it nearly meets the corner of the gutter. The crown is protected by a tongue of asphalt roofing over a flap of aluminum flashing.

Rethinking the Cornice Return

Standard ogee gutters can mimic this classical detail

by Scott McBride

Construction of traditional cornice returns (top photo, p. 136) is a lost art. Like many of the finer points of residential carpentry, returns were cast aside after WWII in the rush toward mass-produced housing. They were replaced by the "pork chop," or ear board, a no-nonsense way of resolving the transition from rake board to eaves fascia. The pork chop is an unadorned triangular piece of pine tacked up flush with the rake board (bottom right photo, p. 136).

A generation of carpenters matured in the 1960s and 1970s with no other return in their repertoire. Then the post-modern movement came along, and suddenly the classic cornice return had to be reinvented. Unfortunately, there weren't many old-timers left who remembered how to build one (for Bob Syvanen's treatment of cornice returns, see the article on pp. 130-133). The task was complicated further because guttering practices had changed in the interim.

In the 18th and 19th centuries, gutters usually were built into the roof and lined with sheet metal. With the gutter concealed, the cornice was free to return neatly onto the gable. Unfortunately, when concealed gutters leak, the water drips into the cornice or, worse, into the wall. In the late 1800s, builders started switching to metal half-round suspended gutters. But these gutters not only look bland, they also throw the crown molding—the crowning glory of the cornice—into

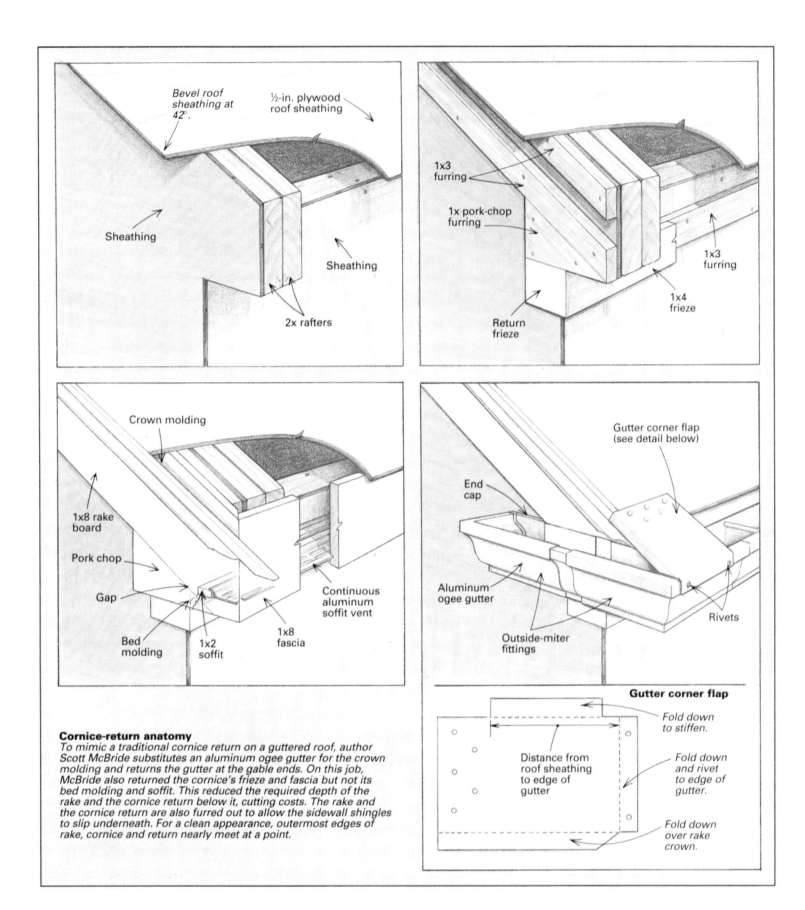

Bevel roof sheathing at 42°.

½-in. plywood roof sheathing

Sheathing

Sheathing

2x rafters

1x3 furring

1x pork-chop furring

1x3 furring

1x4 frieze

Return frieze

Crown molding

1x8 rake board

Pork chop

Gap

Bed molding

1x2 soffit

1x8 fascia

Continuous aluminum soffit vent

Gutter corner flap (see detail below)

End cap

Aluminum ogee gutter

Outside-miter fittings

Rivets

Cornice-return anatomy
To mimic a traditional cornice return on a guttered roof, author Scott McBride substitutes an aluminum ogee gutter for the crown molding and returns the gutter at the gable ends. On this job, McBride also returned the cornice's frieze and fascia but not its bed molding and soffit. This reduced the required depth of the rake and the cornice return below it, cutting costs. The rake and the cornice return are also furred out to allow the sidewall shingles to slip underneath. For a clean appearance, outermost edges of rake, cornice and return nearly meet at a point.

Gutter corner flap

Fold down to stiffen.

Distance from roof sheathing to edge of gutter

Fold down and rivet to edge of gutter.

Fold down over rake crown.

shadow (bottom left photo, p. 136). The metal ogee gutter, or K gutter as it's sometimes called, seems to have appeared in the 1930s. It's an ingenious idea; combine the elegant look of a traditional crown molding with the practicality of a surface-mounted gutter. Given that the strong, easy-to-install ogee gutter is here to stay, I've long sought a way to incorporate it gracefully into a traditional cornice return. After several false starts, I think I've succeeded.

Returning the gutter—My first attempt to integrate the ogee gutter into a traditional return consisted of simply mitering the end of the gutter. It wrapped around the corner over the pork chop and then terminated at an end cap. This looked okay, but it sure didn't sing like the originals.

Then I realized that wooden cornice returns don't just stop once they turn the corner—they turn *another* corner and dead-end at the gable wall. Okay, so I connected a second outside

miter fitting to the first one, with a short length of gutter in between. One leg of this second fitting had to be cut off, leaving just enough room for an end cap to be riveted on. Visually, this helped quite a bit, but there was still something wrong. After studying a lot of old-style cornice returns, it hit me: The outermost edges of the rake, the cornice and the cornice return should all meet at one point, or at least appear that way from the ground. The eye can't help but follow these

lines, and if they converge neatly, a little visual ping goes off.

One problem with my gutter return was that the lip of the eaves gutter sat above the projected roof plane. When viewed from the gable end, the gutter jutted above the rake line. Another problem was the way the return looked from the front. It stuck out 4 in. past the flat rake board like a cow's ear. No wonder it looked clunky.

Working out the details—On a recent job in Rye, New York, I was determined to get my cornice act down. I started drawing the cornice details before the second-floor wall framing was up.

I needed three drawings: a vertical section through the eaves, a section through the rake cut perpendicular to the roof and a horizontal section of the eaves-end cornice return, which is a sort of plan view. My goal was to make the total projection of the eaves, including the gutter, equal to the projection of both the cornice return and the rake. This would allow all three edges to come to a point.

I started with the eaves section, drawing a 2x6 wall with a 2x10 rafter on top of it at the given pitch. Then I sliced thin sections of the different trim elements to use as templates and juggled them around to find a workable configuration. I knew that extending the rake would be the most difficult part; to keep eaves and rake projection roughly equal, I designed the eaves with a minimum overhang.

To keep the eaves overhang as shallow as possible, I called for a narrow 1x2 soffit and slipped the fin of the continuous aluminum soffit vent into a groove plowed in the back of the fascia (drawing, p. 135). The fascia itself would be 7½ in. wide, allowing me to use 1x8 stock for it without ripping. The wider-than-usual fascia was also part of my design strategy; it would allow me to align the outside lip of the gutter with the plane of the roof so that the top edge of the gutter would meet the line of the rake at the corner.

Returning all of the cornice elements onto the gable, minimal as they were, would produce an 11¼-in. overhang for the cornice return. To align the rake with the corner, I would then have to frame an 11¼-in. rake overhang—more work than I had contracted for. I compromised by returning only the fascia and the frieze, in addition to the gutter (photo, p. 134). This would keep the lip of the gutter return close to the house so that it would nearly align with the lip of a 5½-in. crown molding mounted on a furred-out rake board.

A section drawing of the rake helped me figure out the overhang needed on the roof sheathing to support the top of the rake crown molding. The drawing also made it easy to determine the cutting bevel along the edge of the sheathing, which depends on the particular crown used. During installation we let the plywood run long, then snapped a chalkline at the correct distance and cut it with a circular saw set at 42°.

Corner flaps—To bring the rake crown molding down to the corner of the gutter without exposing the molding to the weather, I took my cue from the built-in gutters I've seen. They typically have a narrow tongue of roofing at each end that

Cornice evolution. **Traditional cornice returns often featured a soffit and a full complement of cornice moldings capped by a flashed water table that prevented water damage. In the example above, the top edges of the rake, the cornice and the return meet at a point for an uncluttered look. In the late 1800s, half-round metal gutters began to supplant less-dependable built-in gutters. Bland in appearance, the half-round gutters obscured crown moldings (photo below). Nowadays, standard ogee gutter routinely substitutes for crown molding and simply dead-ends at the corners of the roof. The pork chop, an unadorned 1x triangle that's installed flush with the rake board (photo right), has virtually replaced the traditional cornice return.**

Photos this page: Bruce Greenlaw

extends down to cover the lower end of the rake crown molding. The built-in gutter dead-ends at the inboard edge of the tongue. Around the corner, a flashed water table caps the cornice return's crown molding (the water table is simply a piece of wood that's sloped away from the house to shed water). The rake crown dies into the top of the water table.

I mimicked this arrangement. Instead of extending the roof sheathing to support the tongue of asphalt roofing, though, I installed a simple flap made from white aluminum coil stock (coil stock is trade jargon for prefinished aluminum on a roll). The edges of the flap were turned down with hand seamers to stiffen them, and a fin in front laps over the lip of the gutter. But as it turned out, the gutter used by my gutter contractor is wider than the sample I had obtained from a lumberyard. As a result the flaps fell short, but they still get enough support from the extended rake crown. The tongues of the asphalt roofing were glued to the flaps with roofing mastic.

Splitless nails—Because of all my planning, installation of the cornices went like clockwork. I set up a cutting bench with a Hitachi 15-in. miter saw recessed into the top. This machine swings to 58°, so I could even cut the acute miter at the bottom of the rake crown molding.

The job's general contractor, Frank DiGiacomo, showed me an exterior trim nail that I had not seen before. Called a "splitless," the nail is made by the W. H. Maze Co. (100 Church St., Peru, Ill. 61354; 815-223-8290). A splitless nail is a cross between a finish nail and a box nail. The small, flat head makes it easier to drive than a finish nail and resists trim warpage. The narrow diameter of the shank is less likely to split wood, which is how the nail got its name. The splitless nail has a hot-dipped zinc coating for corrosion resistance and is available in 6d, 8d and 10d sizes. I used these nails exclusively for assembling the cornice.

A variation at the entry—On the entry porch of the Rye project, where the return would be more visible, I decided to frame a full rake overhang by notching 2x4 lookouts into the end rafters (drawing right). This full overhang allowed me to maintain a constant soffit width for the eaves and the cornice return, the latter of which would line up with the rake crown (photo right). Because there is no gable wall for the cornices to return to, the returns make one extra 90° bend and dead-end into the porch ceiling.

I had intended to use aluminum gutters in place of crown molding for the cornice returns as I had done on the higher gables. But once the gutter was installed, it looked way too wide. I replaced it with a 4½-in. crown molding capped by a flashed water table. This looks good but leaves the rake crown sticking a bit proud of the cornice. It isn't too noticeable, though. When I lived in New York we used to say, "Fugetaboutit." Here in Virginia, they say, "It is what it is." □

Scott McBride is contributing editor of Fine Homebuilding *and lives in Sperryville, Va. Photos by the author except where noted.*

Return of the cornice. **Because the cornice return on this entry porch is so prominent, the author framed a full rake overhang, which maintains a constant soffit width for the eaves and the cornice return. The drawing below shows how the rake overhang is built.**

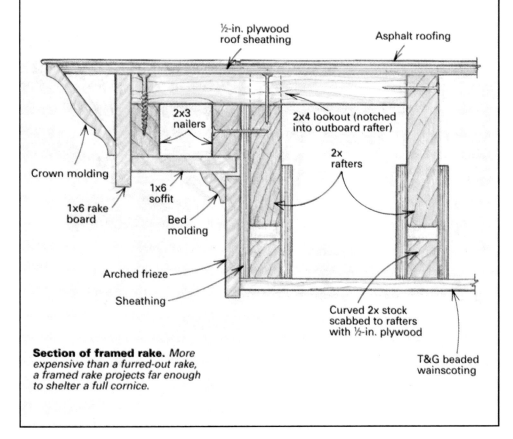

Section of framed rake. *More expensive than a furred-out rake, a framed rake projects far enough to shelter a full cornice.*

½-in. plywood roof sheathing

Asphalt roofing

2x3 nailers

2x4 lookout (notched into outboard rafter)

2x rafters

Crown molding

1x6 soffit

1x6 rake board

Bed molding

Arched frieze

Sheathing

Curved 2x stock scabbed to rafters with ½-in. plywood

T&G beaded wainscoting

Framing with Trigonometry

Getting to know your scientific calculator will make many construction problems easier

by Edwin Zurawski

When I moved from Hawaii to California a few summers ago to build houses with a friend, I learned something that changed my ideas about how to frame a house. Like most carpenters, I knew how to use rafter tables to find the lengths of common and hip rafters for given slopes and spans. I knew how to step off rafter lengths with the framing square. And I had done my share of direct field measurements, from top plate to ridge. But my friend opened the door to a whole new approach by introducing me to trigonometric framing with a calculator.

Trigonometry is the branch of mathematics that deals with the ratios between the sides of a right triangle (any triangle with one 90° angle). It was first documented around 600 B.C. when Thales of Miletus, a Greek, was visiting with mathematicians in Egypt. He used the principles of trigonometry and the length of his shadow to determine the height of the Pyramids. Fifty years later, Pythagoras developed the equation $a^2 + b^2 = c^2$, describing the relationship between the sides and hypotenuse of a right triangle. But while the Pythagorean theorem is limited to calculating the lengths of a triangle's sides, trigonometry can be used to find degrees of angles as well.

What impressed me most about the trigonometric framing method was how ingenious, yet simple, it is to use. You need to learn only two ratios, tangent and cosine. Every carpenter already knows what tangent is but calls it pitch, or rise/run. Tangent is the ratio between the two sides of a right triangle that form the 90° angle. Cosine is the ratio between one of these sides and the hypotenuse. These ratios are illustrated in drawing A (below left).

Trigonometry has become a more accessible tool for builders because scientific calculators are now so inexpensive that you can get one for $15 or $20. I use a Sharp EL-506P, but any make or model will do as long as it has trigonometric functions (photo below). In the following examples, I've rounded off the numbers to simplify the calculations.

A. Trigonometric basics

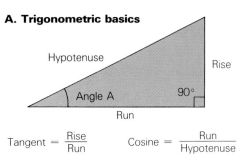

$$\text{Tangent} = \frac{\text{Rise}}{\text{Run}} \qquad \text{Cosine} = \frac{\text{Run}}{\text{Hypotenuse}}$$

Inverse tangent (tan⁻¹) converts tangent to degree. of angle A.

Converting decimals—Since calculators compute in decimals and most building measurements are in feet, inches and fractions of inches, you need to know how to convert from decimal measurements to fractions and vice versa. If you want to convert 6.72 ft. to fractions, for instance, first subtract 6 (the number of whole feet), leaving 0.72 on the calculator screen. Then multiply 0.72 times 12, which converts feet to inches: [.72 × 12 =] 8.64 in. Subtract 8 (the number of whole inches), leaving 0.64 on the screen. Multiply 0.64 times 16 to convert 0.64 in. to sixteenths: [.64 × 16 =] 10.24. Hence 0.64 in. equals approximately ¹⁰⁄₁₆, or ⅝ in. So 6.72 ft. = 6 ft. 8⅝ in.

To convert fractions to decimals, start with the smallest fraction. For example, if you want to convert 12 ft. 5⁵⁄₁₆ in. to decimals, first divide 5 by 16. [5 ÷ 16 =] 0.31 in. Then divide 0.31 in. by 12 to convert it to feet. [.31 ÷ 12 =] 0.026 ft. To convert 5 in. to feet, divide 5 by 12. [5 ÷ 12 =] 0.4167 ft.; [0.4167 + 0.026 =] 0.4427 ft.; and 0.4427 ft. plus the original 12 ft. equals 12.4427 ft.

Calculating common rafters—In order to use trigonometry as a building tool, you have to visualize the right triangles contained within your framework. Once you've done this, if you know the length of one side and the size of one angle, you can use this information to find the remaining dimensions and angles of the triangle. Suppose you're framing a roof with a 6-in-12 pitch on a house that's 30 ft. wide. The length of a common rafter is the hypotenuse of the triangle formed by the ridge, the exterior wall, and the center of the 30-ft. span (drawing B, next page). Here's how to calculate the length of that rafter with trigonometric equations.

On the calculator press [6 ÷ 12 =], and 0.5 appears. That number, rise divided by run, is the trigonometric function known as tangent. You then press [2nd F tan⁻¹] to convert tangent to degrees, and 26.56 appears. The 2nd F key stands for second function and works like the shift key

B. Computing common-rafter length

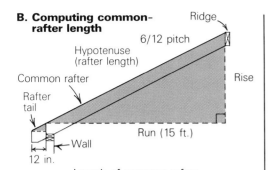

Length of common rafter
$6 \div 12 = 0.5$ (2ndF) (tan⁻¹) 26.56 (cos) 0.8944
$15 \div 0.8944 = 16.77$ or 16 ft. 9¼ in.

Length of rafter tail
$6 \div 12 = 0.5$ (2ndF) (tan⁻¹) 26.56 (cos) 0.8944
$12 \div 0.8944 = 13.42$ or 13⁷⁄₁₆ in.

on a typewriter, allowing some of the keys, tangent in this case, to serve two functions. Tan⁻¹ is called inverse tangent and gives a degree value to a specific tangent. This degree value of tangent can also be used later to make the seat cut of the bird's mouth.

To find the hypotenuse of the triangle, you use run ÷ cosine = hypotenuse. With 26.56 still on the calculator's screen, press [cos] and 0.8944 appears. This is the cosine of 26.56°. Now press [x→m], which stores the cosine of the 6-in-12 triangle in the calculator's memory. Half the span of the 30-ft. house (15 ft.) is the run of the common rafter, so press [15 ÷ RM (recall memory) =] 16.77 ft. (or 16 ft. 9¼ in). This is the distance from the center of the ridge to the plumb cut at the bird's-mouth cut. The actual rafter of course will have to be shortened by half the thickness of the ridge board. (For more on cutting rafters, see pp. 72-77.)

Rafter tails—If the fascia and soffit detail of your house calls for a 12-in. overhang, then you have to visualize another triangle to calculate the additional rafter length (the hypotenuse of that triangle (drawing B). Begin as you did before, by calculating the cosine of a 6-in-12 pitch. [6 ÷ 12 = 0.5 2ndF tan⁻¹ (26.56) cos (.8944) x→m] to store it. Now divide 12 in. (the length of the overhang) by [RM], press [=], and you get 13.42 in. You would therefore add 13⁷⁄₁₆ in. to the length of your common rafter.

Hip rafters—Since the span of the house is 30 ft., then the level run of the hip rafter is the hypotenuse of a 15-ft. by 15-ft. right triangle (drawing C). On a right triangle with two equal

C. Run of the hip rafter

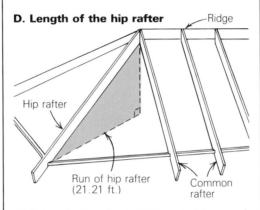

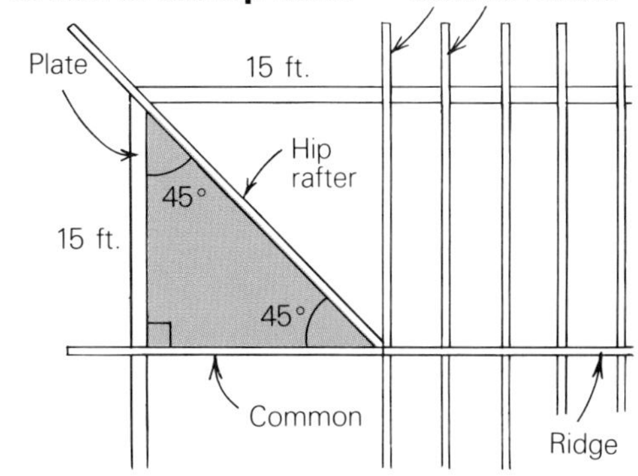

45 (cos) 0.7071
$15 \div 0.7071 = 21.21$, or 21 ft. 2½ in.

sides, an isosceles right triangle, you already know from your high-school geometry that the angle between either of these sides and the hypotenuse is 45°. To find the hypotenuse of a triangle with two equal sides, you simply press [45], then [cos] and 0.7071 appears. Then press [x→m] to store the cosine in the calculator's memory. Now clear the screen and press [15 ÷ RM =] 21.21 ft. This is the level run of the hip.

Hip rafters rise the same distance as common rafters but they have a longer run, since they're cutting diagonally across the building, and therefore they have a different pitch. Instead of 12, the unit run of hip rafters is 17, which is the rounded-off length of the hypotenuse of an isosceles right triangle with 12-in. sides (for more on this, see pp. 72-77).

To find the length of the hip rafter in this example (drawing D), you divide 21.21 ft., the run of the hip, by the cosine of a 6-in-17 triangle. First press [6 ÷ 17 =] and 0.3529 appears. Press [2nd F tan⁻¹] and you get 19.44°. Next hit [cos] and 0.943 appears. Press [x→m] and the

D. Length of the hip rafter

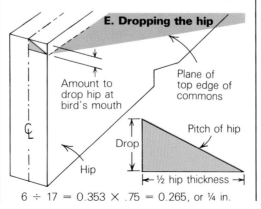

Unit rise of a hip rafter is 17 ft., so in this example the pitch of the hip is 6/17.
$6 \div 17 = 0.353$ (2ndF) (tan⁻¹) 19.44 (cos) 0.943
$21.21 \div 0.943 = 22.49$, or 22 ft. 5⅞ in.

cosine of a 6-in-17 slope is now stored in the calculator's memory. Now clear the screen and press [21.21 ÷ RM =] 22.49 ft. (22 ft. 5⅞ in.) This is the length of the hip rafter.

The hip rafter has to be shortened by a distance equal to half the 45° thickness of the ridge for a single cheek cut or half the 45° thickness of the common rafter for a double cheek cut. Also, the hip must be dropped at the seat cut in order to bring it into alignment with the common rafters on either side of it. Here is the formula for calculating how far to drop the hip rafter: rise divided by 17, multiplied by half the thickness of the hip. As shown in drawing E,

E. Dropping the hip

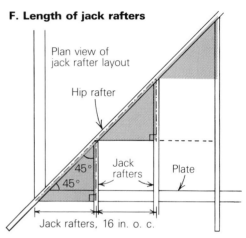

$6 \div 17 = 0.353 \times .75 = 0.265$, or ¼ in.

press [6 ÷ 17 = 0.3529 × .75 (¾ in.—half the thickness of a 1½-in. hip rafter) =] 0.2647 in. I convert 0.2647 in. into sixteenths by multiplying it by 16. [.2647 x 16 =] 4.235 (⁴⁄₁₆) or ¼. You drop the hip ¼ in. for a 6-in-17 slope, if the hip rafter is 1½ in. thick.

Jack rafters—Jack rafters have the same pitch as the common rafters. If you start your jack-rafter layout at the corner of the building, then the on-center spacing of the jack rafters is also the run of the first jack rafter. In drawing F, the first jack is 16 in. o. c. from the corner of the building, and so you find the length of the jack by dividing the run (16 in.) by the cosine of a 6-in-12 pitch.

Here's how to find the length of the first jack rafter. On the calculator, press [6 ÷ 12 =] (0.5

F. Length of jack rafters

Plan view of jack rafter layout

Hip rafter

Jack rafters

Plate

Jack rafters, 16 in. o. c.

$6 \div 12 = 0.5$ (2ndF) (tan⁻¹) 26.56 (cos) 0.8944
$16 \div 0.8944 = 17.89$, or 17⅞ in.

appears), then [2nd F tan⁻¹] (26.56° appears), then [cos] (0.8944 appears), and finally [x→m]. The cosine of a 6-in-12 pitch is now stored in the calculator's memory. Next clear the screen and press [16 (16 in.) ÷ RM =] 17.89 in. (17⅞ in.). Shorten the jack rafter by half the 45° thickness of the hip rafter (this will be 1¹⁄₁₆ in. for a hip that is 1½ in. thick). Then add 17⅞ in. to each succeeding jack rafter. Note that these jack lengths are the distance along the center of the rafter.

I've described methods for hip and jack calculations as they apply to 45° hips. These methods will work for odd-angle hips, but their application is more complicated. Just remember, you still have to break the framing down visually into right triangles.

Trigonometry and scientific calculators liberate you from rafter tables and help you develop skills that make complex problems seem easy. These methods that I've described to calculate rafter lengths will also work with stairs, or with any construction problem involving angles.

If you spend a little time with tangent and cosine, before long you'll be able to do the calculating at home, so you'll be ready to build when you get to work. Anyone curious about a better way to build and willing to spend $15 for a scientific calculator should give it a try.   □

Edwin Zurawski is a general contractor in Concord, Calif.

A Glossary of Roofing Terms

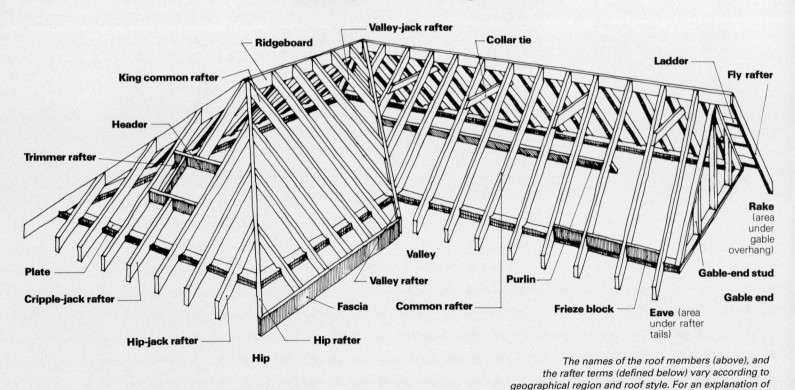

The names of the roof members (above), and the rafter terms (defined below) vary according to geographical region and roof style. For an explanation of how hip and gable roofs are framed, see the article on pp. 72-77.

Span—the horizontal distance between the outside edges of the top plates.

Rise—the vertical distance measured from the wall's top plate to the intersection of the pitch line and the center of the ridge.

Run—the horizontal distance between the outside edge of the top plate and the center of the ridge; in most cases, half the span.

Slope—a measurement of the incline of a roof, the ratio of rise to run. It is typically expressed using 12 as the constant run.

Pitch—has become synonymous with slope in modern trade parlance. It is actually the ratio of the rise to the span. A roof with a 24-ft. span and a rise of 8 ft. has a 1-to-3 pitch. Its slope is 8 in 12. Two ways of saying the same thing.

Unit rise—the number of inches of rise per foot of run.

Unit run—this distance is always 12 in.

Common difference—the difference between the length of a jack rafter and its nearest neighboring jack on a regular hip or valley when they are spaced evenly. This is also the same measurement as the length of the first, or shortest, jack.

Rafter pattern—a full-scale rafter template used to mark the other rafters for cutting. It can be tried in place for fit before cutting all the rafters.

Layout tee—a short template cut from the same stock as the rafters and used for scribing repetitive plumb cuts, tail cuts and bird's mouths.

Tail—the part of a rafter that extends beyond the heel cut of a bird's mouth to form the overhang or eave.

Pitch line—an imaginary line, also called the **measuring line**, that runs parallel to the rafter edges at the height of the full depth of the heel cut on the bird's mouth. In common practice, rafters are measured along their bottom edge.

Theoretical length—the length of a rafter without making allowances for the tail or ridge reduction. Also called the **unadjusted length**.

Bird's mouth—also called a **rafter seat**. It is the notch cut in a rafter that lets it sit on the double plate. It is formed by the plumb heel cut and the seat cut, which is a level line.

Plumb cut—any cut that is vertical when the rafter is in position on the roof. Also used as a reference to the top cut on a rafter where it meets the ridgeboard.

Level cut—any cut that is horizontal when the rafter is in position on the roof.

Tail cut—the cut at the outer end of the rafter. If cut at the outside edge of the double plate, it is a flush cut. All the other traditional tail cuts let the rafter overhang the plates—**heel cut** (level), **plumb cut** (vertical), **square cut** (perpendicular to the length of the rafter) or **combination** level and plumb cuts.

Side cut—also called a **cheek cut**, is the compound angle required for the proper fitting of roof members that meet in an intersection of less than 90°, and other than level. This applies to jacks that connect with hips and valleys.

Ridge reduction—rafter lengths are calculated to the center of the ridge of the roof. This doesn't take into account the thickness of the ridgeboard. This allowance reduces the theoretical length of the rafter by one-half the thickness of the ridgeboard. The layout line drawn parallel to the plumb cut that represents this allowance is called the **shortening line**.

Dropping a hip—the amount by which the bird's mouth on a hip rafter must be deepened to allow the top of the rafter to lie in the same plane as the jack and common rafters. This ensures that the roof sheathing will nail flat without having to bevel the top edges of the hip, a process known as **backing**. —*P.S.*

Illustrations: Frances Boynton

INDEX

The articles in this book originally appeared in *Fine Homebuilding* magazine. The date of first publication, issue number and page numbers for each article are given below.